Springer Tracts in Modern Physics
Volume 126

Editor: G. Höhler
Associate Editor: E. A. Niekisch

Springer Tracts in Modern Physics

* denotes a volume which contains a Classified Index starting from Volume 36

Helmut Dosch

Critical Phenomena at Surfaces and Interfaces

Evanescent X-Ray and Neutron Scattering

With 69 Figures

Springer-Verlag Berlin Heidelberg GmbH

Dr. Helmut Dosch

Sektion Physik, Universität München, Geschwister-Scholl-Platz 1
W-8000 München 22, Fed. Rep. of Germany

Manuscripts for publication should be addressed to:

Gerhard Höhler

Institut für Theorctische Teilchenphysik der Universität Karlsruhe, Postfach 69 80,
W-7500 Karlsruhe 1, Fed. Rep. of Germany

*Proofs and all correspondence concerning papers in the process of publication
should be addressed to*

Ernst A. Niekisch

Haubourdinstrasse 6, W-5170 Jülich 1, Fed. Rep. of Germany

Library of Congress Cataloging-in-Publication Data. Dosch, Helmut, 1955–. Critical Phenomena
at Surfaces and Interfaces: evanescent x-ray and neutron scattering Helmut Dosch. p. cm. –
(Springer tracts in modern physics: 126). Includes bibliographical references and index.
1. Surfaces (Physics) – Optical properties. 2. Critical phenomena
(Physics). 3. X-rays–Scattering. 4. Neutrons–Scattering. 5. Grazing incidence. I. Title. II. Series.
QC 1.S797 vol. 126 [QC173.4.S94] 530 s–dc20 [530.4'27] 91-37384

ISBN 978-3-662-14975-1   ISBN 978-3-540-38456-4 (eBook)
DOI 10.1007/978-3-540-38456-4

© Springer-Verlag Berlin Heidelberg 1992
Originally published by Springer-Verlag Berlin Heidelberg New York in 1992.
Softcover reprint of the hardcover 1st edition 1992

Typesetting: Springer T$_E$X in-house system
Production Editor: P. Treiber

57/3140-543210 – Printed on acid-free paper

Preface

In this article I review the scheme of grazing angle x-ray and neutron scattering and in particular its application to critical phenomena at surfaces. After an introductory discussion of elementary x-ray optics and the theoretical concepts to describe the scattering of evanescent x-rays, the scattering phenomena which occur at *real* solid surfaces are treated in detail, for example, how surface roughness, surface waviness, the miscut of the surface or an amorphous surface layer affects the kinematic Bragg scattering intensity.

Apart from the nuclear interaction with matter, neutrons exhibit a strong magnetic interaction. This leads to interesting phenomena in neutron optics which are already exploited nowadays, and which on the other hand can be used for novel neutron scattering experiments at magnetic interfaces. The potential of this new method to obtain microscopic information from the surface of ferromagnets, antiferromagnets, superconductors and from buried ("internal") interfaces is illustrated by recent experiments.

The main part of this book deals with experiments on surface critical phenomena. Here the advantages of x-ray and neutron scattering, which have so successfully been exploited in bulk experiments for many decades, can be used to study surface-related correlation functions in an intriguingly direct way. In the last few years grazing angle scattering of x-rays and (slowly starting, though) neutrons has been providing a lot of exciting details about surface melting, surface roughening, surface-related critical exponents in semi-infinite critical systems and other surface-induced critical phenomena.

In writing this review I intended to address the reader who wants a readable and also comprehensive introduction to the present state of the art of this still rapidly developing research field, but also, and primarily, the experimentalist working with x-rays or neutrons. For this reason I included, for example, a collection of "engineering formulae" related to surface scattering which turned out to be quite helpful to me in "laboratory life". Most of the evanescent scattering cross sections which occur in this article are explicitly derived. (A summary of integrals actually used is given in a separate appendix.) Thus, most parts of this book may also be useful for tutorial courses.

Munich, March 1992   *Helmut Dosch*

Acknowledgements

I am indebted to Klaus Al Usta, Thomas Höfer, Robert L. Johnson (Universität Hamburg), Andreas Lied, Lutz Mailänder, Johann Peisl, and Harald Reichert for the pleasant collaboration within the Munich "surface critical phenomena" group, and to Siegfried Dietrich, Gerhard Gompper, Reinhard Lipowsky, and Herbert Wagner for many inspiring discussions. I profited from the expertise of Jens Als-Nielsen (Risø Nat. Laboratory), Neil D. Ashcroft (Cornell University), Bob Batterman (Cornell University), Jakob Bohr (Risø Nat. Laboratory), Bruno Dorner (Institut Laue Langevin, Grenoble), Robert Feidenhans'l (Risø Nat. Laboratory), Gian Felcher (Argonne Nat. Laboratory), Francois Grey (Risø Nat. Laboratory), Simon Moss (University of Houston), Jeff Penfold (ISIS, RAL) and many others. The technical support and the hospitality at HASYLAB (Hamburg), at the ILL (Grenoble), at ANL (Argonne) and at ISIS (Didcot) during our experiments is greatly appreciated.

Contents

List of Acronyms

ANL	Argonne National Laboratory, Argonne IL (USA)
CRISP	Critical reflection spectrometer
DG	Discrete Gaussian
DWBA	Distorted wave Born approximation
ESRF	European Synchrotron Radiation Facility, F–38042 Grenoble Cedex (France)
EVA	Evanescent neutron wave diffractometer
GID	Grazing incidence diffraction
HASYLAB	Hamburg Synchrotron Laboratory, W–2000 Hamburg 52 (Germany)
HFR	High flux reactor
ICISS	Impact collision ion scattering spectroscopy
ILL	Institut Laue–Langevin, F–38042 Grenoble Cedex (France)
IPNS	Intense Pulsed Neutron Source (spallation source at ANL, USA)
ISIS	Spallation source of the RAL at Didcot, Oxfordshire (UK)
LEED	Low-energy electron diffraction
LEIS	Low-energy ion scattering
LRO	Long range order
MC	Monte Carlo
MF(A)	Mean field (approximation)
ML	Monolayer
NN	Nearest neighbour
NNN	Next nearest neighbour
NSLS	National Synchrotron Light Source, Brookhaven NY (USA)
OP	Order Parameter
PG	Pyrolytic graphite
POSY	Polarized neutron reflectometer
PSD	Position sensitive detector
QSD	Quenched surface disorder
RAL	Rutherford Appleton Laboratory, Didcot, Oxfordshire (UK)
RG	Renormalization group
SID	Surface induced disorder
SIO	Surface induced order
SL	Superlattice
SPLEED	Spin polarized LEED

SRO	Short range order
TOF	Time-of-flight
TOREMA	Total reflection machine
TR	Truncation rod
vdW	van der Waals
XPS	X-ray photospectroscopy

1. Introduction

When a physicist is asked to define the surface of a crystalline solid he presumably would first swallow two aspirins and eventually answer that this object "surface" comprises the first few (say three) atomic layers of the solid that differ significantly in the atomic and electronic structure from the bulk, eventually including foreign atoms from the gas phase adsorbed to it, a definition which actually hits quite well the typical playground of the surface scientist. Exponentially increasing experimental efforts took place in this field in the last two decades. Certainly this is closely related to the advances in the development of surface preparation techniques and microscopic surface probes, such as low-energy electron diffraction (LEED), low-energy ion scattering (LEIS), atom scattering, grazing-incidence diffraction (GID) of x-rays and spectroscopic methods. In all of these studies the surface is regarded as a (quasi-) 2-dimensional object or as a template, where "*2d-physics*" can be investigated.

The surface of a solid, however, can also be considered from a very different view point, namely as the border of a semi-infinite 3-dimensional crystal. Since one knows that perfect order in solid matter can only exist at $T = 0 \, \text{K}$ and in infinite systems, it is pertinent to adress the question, how the fundamental electronic, structural and thermodynamic properties of the solid are altered close to its border. In fact, in the course of the study of phase transitions in socalled semi-infinite condensed matter, it turned out that many fascinating new phenomena may occur. Figure 1.1 should give a flavour of the richness of surface phenomena which may be observed at the surface of a semi-infinite solid: (a) a *wetting* effect, where close to a first-order phase transition the high-temperature phase appears in a thin surface layer (most prominent is the *surface melting* phenomenon), (b) new *surface-critical phenomena* at bulk criticality or (c) the Meißner–Ochsenfeld effect, i.e., the penetration of an *external magnet field* into a superconducting solid. Common to all these phenomena is that they are sparked at the surface and dragged into the bulk by the inherent correlation length ξ of the system. So the notation "surface" becomes more and more diffuse the larger ξ grows, finally, when $\xi \to \infty$, the distinction between surface and bulk disappears at all. The study of these surface effects allows one to investigate how the change in the dimension and the truncated translational invariance alters the character of phase transitions.

In our every day life we are almost exclusively faced with the surface of the objects around us and many surface-induced phenomena are of practical and technological importance. I will mention a few examples: As one knows, or-

1

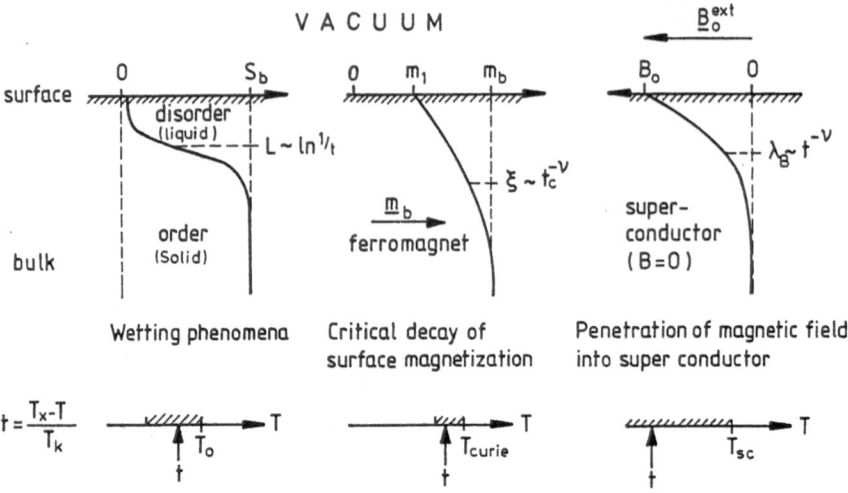

Fig. 1.1. Surface phenomena in semi-infinite solids triggered by bulk phase transitions

dering and precipitation phenomena exert a crucial influence on the engineering properties of two-phase alloys (see e.g. [Warlimont, 1974]), therefore one must expect that surface effects, like surface-induced disordering and surface segregation, modify some surface-related material constants, like tribological properties, in a crucial way. Surface segregation, the surface enrichment of solute atoms, plays a dominant role in corrosion and heterogeneous catalysis. When brazing alloys are considered, their wetting characteristics are among else governed by surface segregation. It was also suggested that the status of the surface may be co-responsible for certain irregularities ("*yield points*") in the stress-strain curve of metals and alloys [Weil et al., 1963; Kramer and Kumar, 1969]. It appears reasonable to assume that the border of a solid may play an important role in various thermodynamic processes, as in nucleation and growth, in martensitic transitions, in spinodal and eutectic decomposition. A systematic investigation of such phenomena on a microscopic level could open a new field of material science which one may call "*Near Surface Material Science*".

At the surface of condensed matter another interesting critical phenomenon occurs, the total external reflection of x-rays and neutrons when the incidence angle with respect to the surface falls short of the associated critical angle. Also this surface phenomenon is not confined to the toplayer, as one might intuitively assume, but rather occurs within a finite surface skin whose thickness is given by the penetration depth of the evanescent wave field. This exponentially fast decay of the x-ray wave field inside the matter (within typically 50 Å) has been very successfully exploited in "*surface crystallography*" of the structures of ultraclean single crystal surfaces and of 2-dimensional adsorbates on crystal surfaces. The weak surface Bragg rods from, say, a surface reconstruction become observable, when the strong bulk scattering is suppressed by the extremely limited penetration depth of evanescent x-ray waves (for a recent review see [Feidenhans'l, 1989]).

To my knowledge Bohr et al. [1985] have solved a surface reconstruction for the first time using this new x-ray technique. The application of grazing angle x-ray scattering to such structurally *"decorated"* surfaces will *not* be touched in this book, here I rather focus onto another point:

In the context of the surface phenomena adressed above it is intriguing that the penetration depth of evanescent x-rays and neutrons is just of the same order of magnitude as a typical correlation length which governs the thickness of the surface-induced phenomena in semi-infinite solids, therefore these exotic waves seem ideally apt for the investigation of the microscopic details of these phenomena which settle between 2 and 3 dimensions. Indeed, considerable theoretical and experimental efforts have been devoted in the last years to exploit the possible applications of the scattering from evanescent wave fields. An important part of this article therefore reviews the state of the art of this young experimental technique: I tried to summarize the relevant theoretical approaches (as critical surface scattering of x-rays, magnetic surface scattering of neutrons and surface roughness induced scattering) and the various x-ray and neutron scattering experiments which have been performed in the attempt to extract microscopic informations from the near-surface regime of a solid. Thereby I have put particular emphasis onto the investigation of semi-infinite phase transitions. The review is organized as follows:

In Chaps. 2 and 3 the grazing angle x-ray and neutron scattering from solid surfaces are discussed. Apart from the different interactions with matter the basic concepts of grazing angle scattering are the same for x-rays and neutrons, therefore the evanescent scattering phenomena discussed in the context of x-rays also hold in most cases for neutrons and vice versa. So I discussed e.g. the evanescent Bragg scattering law in Chap. 2 and the roughness-induced diffuse scattering in Chap. 3, thereby tacitly assuming that it is clear to the reader that they hold for both techniques.

To my knowledge one of the first experimental studies of specular reflection of x-rays appeared 1931 in the *Annalen der Physik* by Kiessig [1931] who earned his "Doktorgrad" at the Technische Hochschule München with *"Untersuchungen zur Totalreflexion von Röntgenstrahlen"* on Ni surfaces using various x-ray wavelengths. By this he could deduce impressively accurate values for the x-ray dispersion in Ni. Already Compton [1922] proposed to measure the x-ray index of refraction by specular reflection. Later Picht [1929] published in his *"Beitrag zur Theorie der Totalreflexion"* (dedicated to Max von Laue on the occasion of his 50th birthday) a detailed theoretical analysis of the energy flow across the totally reflecting interface. Today specular reflectivity studies of surfaces are widely applied, and if it is correct to say that one can call an experimental technique a "standard method" as soon as chemists and biologists are using it, then this surface tool is just becoming standard nowadays. The possibility to exploit the limited penetration depth of evanescent x-ray waves to render x-ray scattering surface-sensitive was first mentioned by Eisenberger and coworkers [Marra et al., 1979; Eisenberger and Marra, 1981]. This clever idea opened the door to many fascinating applications of x-ray scattering to problems related with surface science. Some of them will be adressed in this review.

Neutron reflectivity and neutron scattering under the condition of total external reflection is discussed in Chap. 3. Since the development of neutron sources went along with the *"Manhattan Project"*, it is no big surprise that the first totally reflected neutron beam was reported by Fermi and coworkers [Fermi and Zinn, 1946; Fermi and Marshall, 1946] at the 272nd meeting of the American Physical Society held in June 20–22, 1946 at Chicago. Quoting from the minutes of the meeting, this was *"the first great nuclear-physics meeting at which a large amount of work done under the famed 'Manhattan Project' was finally brought to light"*. The incentive for the reported neutron reflectivity studies (performed at the heavy water pile of the Argonne Laboratory) was the open question, whether or not the neutron suffers a phase shift during the scattering process from the nuclei. The first specular reflection experiment from a magnetic surface has been performed by Hughes and Burgy [1951] again in the context of a very fundamental question, namely the origin of neutron magnetic moment (see Sect. 3.1).

In Chaps. 4 and 5 I tried to give a summary of the theoretical and experimental activities in the application of evanescent x-ray and neutron scattering to the characterization of phase transitions in semi-infinite solids. The investigation of surface-modified critical behaviour at a bulk critical point with x-ray and neutron scattering under glancing angles has been sparked by Dietrich and Wagner [1983] who calculated the scattering cross section of surface critical scattering. The accurate measurement of these intensities gives access to new universal surface exponents. In the last years it turned out that the surface of a system also plays a fundamental role in first order phase transitions: While the thermodynamical phenomenology is well described by Clausius Clapeyron-type equations, there remains the challenge to understand the mechanism for the spontaneous symmetry break which occurs with the sudden appearance of the high temperature phase. Some experiments on various systems give direct evidence that the border of a solid is the natural inhomogeneity which helps the system to reach the disordered state. In this context experiments on the order-disorder transition in Cu_3Au, on surface melting of Pb and Al, and, related to that, on the surface roughening of Ag and Cu will be discussed in detail. Here the potential of evanescent x-ray and neutron scattering becomes particularly evident. In the interpretation of the observed grazing angle scattering signal within the framework of semi-infinite thermodynamics and statistical mechanics I tried to make the relevant results from these theories plausibel. For a rigorous discussion of the foundations of the thermodynamics at the edge of a solid I refer the reader to the contribution by Reinhard Lipowsky in a later volume of this series.

I already mentioned that among all surface-critical phenomena *surface melting* is probably the most exciting surface effect, in particular the surface melting of *ice*, because it touches our everyday life in a very vital way. Ion beam experiments on ice single crystals showed that prior to melting a disordered ("liquid") film appears at the surface [Golecki and Jaccard, 1978]. Its thickness apparently is as large as 40 nm at $-10°$ C and decreases only slowly upon decreasing temperature, until between $-30°$ C and $-40°$ C it seems to disappear. Although these quantitative conclusions on ice are still controversial, there are good reasons to assume that the surface melting effect itself occurs in fact at ice surfaces and could shed a new light on many surface properties of ice. It is e.g. a pertinacious scientific fairy-tale that the *"skating effect"* is due to the pressure-induced melting of ice, however, already a rough estimate of this pressure effect tells one that something must be wrong: When we assume that a typical runner has a length of 25 cm and an effective width of, say (optimistically), 0.5 mm, then the exerted local pressure on the ice surface by a boy weighing 50 kg is $p \simeq 20$ atm. From the latent heat of ice and the solid-liquid density difference it follows (via the Clausius–Clapeyron equation) that the pressure dependence of the solid-liquid coexistence line is given by $(dp/dT)_{sl} = -138$ atm/$°$ C. This, however, would mean that the poor boy would only be able to skat properly as long as the temperature does not drop below $-0.15°$ C (!). The skating effect is apparently more complicated than discussed qualitatively in high school in simple terms of a local pressure underneath the runner. If a liquid layer is necessary at all to reduce the friction between the runner and the ice, it is quite evident that a naturally present quasiliquid surface layer down to, say, $-20°$ C would allow skating even in frostly winters. Since (fortunately) temperatures below $-20°$ C usually do not occur in our hemisphere, we never experienced that skating on ice becomes impossible, but we know the experience of the Arctic explorers. Wright [1924], who participated the *"Scott Polar Expedition 1911–1913"* ([Scott, 1933]: *"The Land that God gave Kain"*) to perform among else detailed measurements of gravity with a precision pependulum, reported:

"Quite apart from any question of the hardness of the snow, however, the surface temperature has an important influence. Our opinion was that the friction decreased steadily as the temperature rose above zero Fahrenheit[1], the presence of brilliant sunlight having an effect, which was more than a psychological one, on the speed of advance. Below zero Fahrenheit the friction seemed to increase progressively as the temperature fell, as if a greater and greater proportion of the friction were due to relative movement between the snow grains and less to sliding friction between runner and snow."

---

[1] which is $-18°$ C

2. Evanescent X-Ray Scattering

2.1 X-Ray Optics

The propagation of an electromagnetic plane wave $E(r) = E_0 \exp(i(k \cdot r - \omega t))$ in a medium characterized by an index of refraction n follows from Maxwell's equations to obey the Helmholtz equation

$$\text{rot rot } E(r) + k^2 n^2 E(r) = 0 \quad , \tag{2.1}$$

where $k = \omega/c$ is the wavevector and ω the frequency of the electromagnetic wave. The index of refraction is given by [Compton and Allison, 1935; James, 1948]

$$n^2 = 1 + \frac{Ne^2}{\varepsilon_0 m} \sum_j \frac{f_j}{\omega_j^2 - \omega^2 - i\omega\gamma_j} \tag{2.2}$$

and thus determined by the electronic resonances ω_j. N is the number of atoms per unit volume, m the electronic mass, γ_j a phenomenological damping factor and f_j the "oscillator strengths" which fulfill the sum rule $\sum_j f_j = Z$ (charge number of the atoms). If $E(r)$ is a plane x-ray wave, then $\omega > \omega_j$ and n degenerates to

$$n(r) = n + \Delta n(r) = 1 - \delta + i\beta + \Delta n(r) \tag{2.3}$$

with

$$\delta = \lambda^2 \frac{Nr_e Z}{2\pi} \quad (N \text{ is the average number density}), \tag{2.4}$$

$$\beta = \lambda \frac{N\sigma_a}{4\pi} = \frac{\lambda\mu}{4\pi} \quad (\text{extinction coefficient}). \tag{2.5}$$

In (2.4,5) λ is the x-ray wavelength, $r_e = e^2/(4\pi\varepsilon_0 mc^2) = 2.814 \times 10^{-5}\,\text{Å}$ the classical electron radius, σ_a the absorption cross section and μ the linear photoabsorption coefficient (see A.7). For later purposes $n(r)$ has been split into the average value n of the electronic jellium and the local deviation $\Delta n(r)$ from it. The phenomenon of total external reflection from the surface of condensed matter occurs, because n is less than 1. By inserting typical numbers into (2.4) and (2.5) one finds that both quantities δ and β are of the order 10^{-5}, thus, the x-ray critical angle

$$\alpha_c = (2\delta)^{1/2} = \lambda \left[\frac{N r_e Z}{\pi}\right]^{1/2} \tag{2.6}$$

is usually of the order of some fractions of a degree (see also A.3).

The effective number of electrons contributing to the jellium is affected by the presence of absorption edges. Denoting $f = f_0 - (\Delta f_1 + i\Delta f_2)$ the form factor of the atom, with $(\Delta f_1 + i\Delta f_2)$ as the dispersion correction, the charge number Z in (2.6) has to be replaced by $Z_{\rm eff} = Z - \Delta f_1$. The K-shell absorption mechanism e.g. in gold can be described as follows [Snell, 1973]:

> Said a K-shell electron in gold,
> "I'm thinking of leaving the fold
> To be hit like a hammer
> By an incoming gamma
> In freedom I'll live till I'm old."

We consider now the situation of a plane x-ray wave in vacuum impinging on a flat surface of a crystal (at $z = 0$) under a shallow angle α_i which is in the vicinity of the critical angle α_c (Fig. 2.1). The linearly polarized incident x-ray wave field $E_i(r) = E_i \exp(ik_i \cdot r)$ reads in the coordinate system of Fig. 2.1:

$$E_i = E_{i\parallel}\begin{pmatrix} \sin\alpha_i \\ 0 \\ -\cos\alpha_i \end{pmatrix} + E_{i\perp}\begin{pmatrix} 0 \\ 1 \\ 0 \end{pmatrix} \quad , \quad k_i = k\begin{pmatrix} \cos\alpha_i \\ 0 \\ \sin\alpha_i \end{pmatrix} \quad , \tag{2.7}$$

where $E_{i\parallel}$ and $E_{i\perp}$ denote the components of the electric field vector parallel and perpendicular to the plane of incidence ($x - z$ plane). The solution of the problem follows from the ansatz

$$E_t(r') = \mathcal{T}_i E_i \, e^{ik'_i \cdot r'} \qquad r' \text{ in halfspace } z \geq 0 \ ,$$

$$E_r(r) = \mathcal{R}_i E_i \, e^{ik_r \cdot r} \qquad r \text{ in halfspace } z < 0 \tag{2.8}$$

for the transmitted and reflected waves fields $E_t(r')$ and $E_r(r)$ with the transmission and reflection tensors

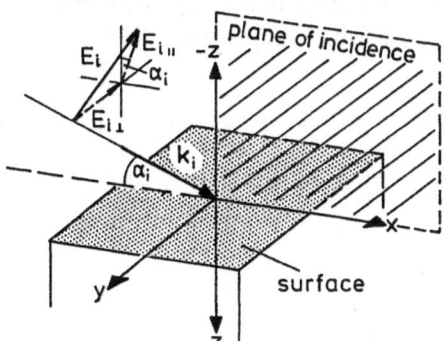

Fig. 2.1. Monochromatic x-ray with wavevector k_i and field vector E_0 impinging onto an interface (at $z = 0$) at a grazing angle α_i

$$\mathcal{T}_i = \begin{pmatrix} T_{ix} & & 0 \\ & T_{iy} & \\ 0 & & T_{iz} \end{pmatrix} \quad , \quad \mathcal{R}_i = \begin{pmatrix} R_{ix} & & 0 \\ & R_{iy} & \\ 0 & & R_{iz} \end{pmatrix} \tag{2.9}$$

and

$$\boldsymbol{k}'_i = k \begin{pmatrix} \cos \alpha_i \\ 0 \\ sinr\ \alpha_i \end{pmatrix} \quad \text{and} \quad \boldsymbol{k}_r = k \begin{pmatrix} \cos \alpha_i \\ 0 \\ -\sin \alpha_i \end{pmatrix} \tag{2.10}$$

as the refracted and reflected waves with the notation $sinr\ \alpha \equiv (\sin^2 \alpha - 2\delta + 2i\beta)^{1/2}$ for the "refracted $\sin \alpha$" in the medium. The Fresnel coefficients $T_{i\varphi}$ and $R_{i\varphi}$ are calculated from (2.1) under the boundary conditions that across the interface the tangential components of the field vectors are continuous [Born and Wolf, 1980]:

$$T_{ix} = \frac{2\ sinr\ \alpha_i}{n^2 \sin \alpha_i + sinr\ \alpha_i} \quad , \quad R_{ix} = -\frac{n^2 \sin \alpha_i - sinr\ \alpha_i}{n^2 \sin \alpha_i + sinr\ \alpha_i}$$

$$T_{iy} = \frac{2 \sin \alpha_i}{\sin \alpha_i + sinr\ \alpha_i} \quad , \quad R_{iy} = \frac{\sin \alpha_i - sinr\ \alpha_i}{\sin \alpha_i + sinr\ \alpha_i} \tag{2.11}$$

$$T_{iz} = \frac{2 \sin \alpha_i}{n^2 \sin \alpha_i + sinr\ \alpha_i} \quad , \quad R_{iz} = \frac{n^2 \sin \alpha_i - sinr\ \alpha_i}{n^2 \sin \alpha_i + sinr\ \alpha_i} \quad .$$

Note that the amplitudes of the transmitted waves are $|\boldsymbol{E}_{t\|}| = |n|\,\boldsymbol{E}_{t\perp} \equiv \boldsymbol{E}_t$ and therefore indistinguishable for any practical case, since $|n| = 0.9999$. The same argument holds for the reflected amplitudes which can be checked by the reader quite simply. (We denote the x-ray reflectivity in the following by R_i).

Figure 2.2 shows $|R_i|^2 \equiv (E_r/E_i)^2$ and $|T_i|^2 \equiv (E_t/E_i)^2$ as a function of the incident angle α_i normalized to α_c for a fixed value of δ and various values of the ratio β/δ. The properties of $|R_i|^2$ have been analyzed in detail by Parratt [1954]. Since we don't focus onto x-ray reflectivity measurements in the discussion of semi-infinite matter, we refer the reader to this excellent early paper. For glancing angle x-ray and neutron scattering the transmitted intensity $|T_i|^2$ will turn out to be most important. The striking feature of $|T_i|^2$ is the enhancement of the transmitted beam at the critical angle which originates from the coherent coupling between the incident, reflected and transmitted x-ray wave fields at the surface. The transmitted wave is exponentially damped into the less dense medium,

$$E_t(\boldsymbol{r}') \propto e^{i\,\text{Re}\{\boldsymbol{k}'_i\}\cdot\boldsymbol{r}'} e^{-z/\ell_i} \tag{2.12}$$

with the penetration depth ℓ_i given by

$$\ell_i \equiv |\text{Im}\{k'_{iz}\}|^{-1} = \frac{\lambda}{2\pi l_i} \quad \text{with}$$

$$l_i = 2^{-1/2}\left\{ (2\delta - \sin^2 \alpha_i) - \left[(\sin^2 \alpha_i - 2\delta)^2 + 4\beta^2 \right]^{1/2} \right\}^{1/2} \tag{2.13}$$

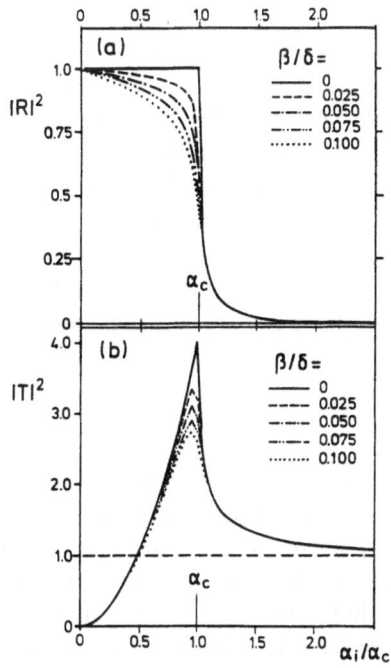

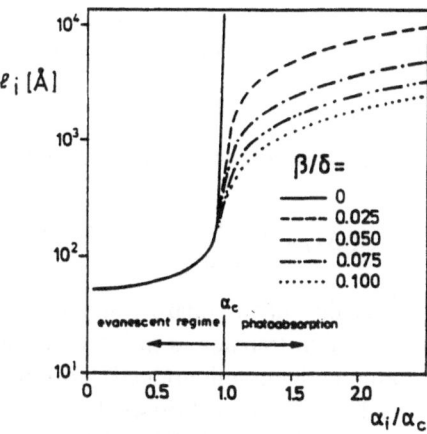

◀ **Fig. 2.2.** (a) X-ray reflectivity $|R_i|^2$ and (b) transmissivity $|T_i|^2$ versus α_i for various β/δ

Fig. 2.3. Penetration depth ℓ_i of evanescent x-rays versus α_i/α_c for various β/δ

which has the asymptotic value $\ell_{i0} = \lambda/2\pi\alpha_c \approx 50\,\text{Å}$ for $\alpha_i \to 0$ ("*evanescent wave*"). The evanescent wave field travels parallel to the surface of an ideally transparent crystal, since in this case $\text{Re}\{k_i'\} = k_{i\parallel}$ ($k_{i\parallel}$ is the wave vector component parallel to the surface). In real systems, however, where $\beta \neq 0$, a small oscillatory component perpendicular to the surface *always* remains and has to be taken into account properly in any detailed quantitative analysis of grazing angle scattering intensities. In Fig. 2.3 the penetration depth ℓ_i of the evanescent ("*distorted*") wave is shown for various values of β/δ. I want to note that this unusual propagation of electromagnetic waves in the less dense medium has nicely been observed by Goos and Haenchen [1947] in the case of visible light ("*Ein neuer und fundamentaler Versuch zur Totalreflexion*").

Total external reflection of x-rays from a "*Fresnel surface*" does not occur in reality no matter how mirror-like and macroscopically flat the crystal surface may be, because x-rays "feel" the atoms which the real surface is built of. This unavoidable microscopic surface roughness caused by the local deviations $\Delta z(\mathbf{r}_\parallel)$ of the surface atoms from the mathematical position $z = 0$ has to be included into the x-ray Fresnel coefficients which then, however, can in general no longer deduced analytically from the Helmholtz equation (2.1).

Surface roughness plays a very important, though negative, role in the application of x-ray total reflection mirrors, as in x-ray astronomy. Here the microscopic roughness of the mirror surfaces limits the resolution of the device[1]. The influence of surface roughness on the x-ray reflectivity has been discussed in the literature [Als–Nielsen et al., 1982; Als–Nielsen, 1986; Croce, 1979; Pershan and Als–Nielsen, 1984], a well-known example hereof is the measurement of the capillary surface waves of water [Braslau et al., 1985]. Within a simple first Born approximation the x-ray reflectivity of a rough surface can be described by (see also [Beckmann and Spizzichino, 1963; Braslau et al., 1987])

$$R_\varrho = R e^{-2k_{iz}^2 \varrho^2} \qquad (2.14)$$

when the roughness $\Delta z(\mathbf{r}_\parallel)$ is statistically distributed with the Gaussian weight

$$w(\Delta z) = \left(\frac{1}{2\pi\varrho^2}\right)^{1/2} e^{-\Delta z^2/2\varrho^2} \qquad (2.15)$$

(ϱ^2 is the mean square roughness). In (2.14) $2k_{iz} \equiv Q_z$ is the perpendicular momentum transfer as observed in the vacuum. It was pointed out in the (to my opinion not enough noticed) work by Croce [1979], Nevot and Croce [1980] and Vidal and Vincent [1984] that the solution (2.14) violates symmetry relations of electrodynamics: We go back to the Helmholtz equation (2.1) which has, in addition to the solution (Fig. 2.4a) which we have considered above, a time reversed solution (Fig. 2.4b). Since the general solution (any linear combination of these two fundamental solutions) is completely symmetric with respect to the interface (Fig. 2.4c), one should expect that any property of the interface should enter the Fresnel coefficients in the same symmetric way independent of the approximations made: The Fresnel transfer matrix which connects all four possible wavefields E_s ($s = 1, \ldots, 4$) is well known from conventional optics [Born and Wolf, 1980] and given by

$$\begin{pmatrix} E_1 \\ E_2 \end{pmatrix} = \begin{pmatrix} F_{11} & F_{12} \\ F_{12}^* & F_{11}^* \end{pmatrix} \cdot \begin{pmatrix} E_3 \\ E_4 \end{pmatrix} \qquad (2.16)$$

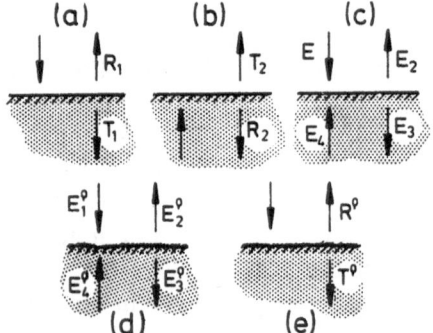

Fig. 2.4a-e. Solution of the Helmholtz equation at an interface (reflected and transmitted waves): (a) "forward" solution, (b) time reversed solution to (a), (c) full symmetric solution, (d) perturbed symmetric solution (rough interface), (e) perturbed "forward" solution (for further explanation see text)

---

[1] In May 1990 the German Röntgen satellite ROSAT furnished with an imaging x-ray telescope based on total reflection mirrors has been deployed into space [Trümper, 1990, 1991]. The surface roughness of the focusing mirrors is as low as 3 Å.

with the coefficients

$$F_{11} = \frac{k_{iz} + k'_{iz}}{2k_{iz}} \, e^{-i(k'_{iz} - k_{iz})z} \quad ,$$

$$F_{12} = \frac{k_{iz} - k'_{iz}}{2k_{iz}} \, e^{i(k'_{iz} + k_{iz})z} \quad . \tag{2.17}$$

Consider now the case when the surface is rough (see Fig. 2.4d): By treating the roughness as a small perturbation of the Fresnel surface any Fresnel amplitude $F_q = A \, \exp(iqz)$ in (2.17) gets decorated by the proper Debye–Waller factor $\exp\big((-1/2)\, \varrho^2 q^2\big)$. This approximation for the solution of the Helmholtz equation at a rough surface preserves the symmetry with respect to time reversal and can now be applied to the situation depicted in Fig. 2.4e which is recovered from Fig. 2.4d by setting $E_4^\varrho = 0$. It is then straightforward to deduce the associated reflection coefficient to be

$$R_\varrho = \frac{F_{12}^*}{F_{11}} = R \, e^{-2k_{iz} k'_{iz} \varrho^2} \quad . \tag{2.18}$$

When the incidence angle α_i becomes steeper, k'_i merges into k_i and the two expressions (2.14) and (2.18) for the x-ray reflectivity R_ϱ become identical. On the other hand, it naturally follows from (2.18) that $|R_\varrho|^2 = 1$ for $\alpha_i \leq \alpha_c$ independent of the atomic roughness of the surface, thus, the expression (2.18) gets rid of the undefined Debye–Waller factor (2.14) below the critical angle. The Debye–Waller factor induced loss of reflectivity manifests the outcrop of a diffuse scattering evoked by the microscopic irregularities of the interface. By studying this roughness-induced diffuse intensity [Andrews and Cowley, 1985; Sinha et al., 1988] one may win detailed information on the height-height correlations (see Sect. 3.3.1). For the further discussion of grazing angle scattering we will require the transmission coefficient of a rough surface. We find

$$T_\varrho = \frac{1}{F_{11}} = T \, e^{\frac{1}{2}(k'_{iz} - k_{iz})^2 \varrho^2} \quad , \tag{2.19}$$

i.e. the transmission function gets *enhanced* by surface roughness within the Nevot–Croce theory. This has apparently a very intuitive reason: The rougher the surface the less it reflects x-rays and the more it allows x-rays to be transmitted into the less dense medium. (Within the simple first Born approximation T_ϱ would be decreased by surface roughness.)

I want to end the discussion of x-ray optics with a short note on one of the most pretentious experimental efforts exploiting the total external reflection phenomenon, namely the feasibility and performance studies of *"x-ray guide tubes"*: The x-ray-*"guide effect"* of untapered capillaries has been demonstrated by Mosher and Stephanakis [1976] and Vetterling and Pound [1976] using x-ray energies of $E = 5.8$ keV and 14.4 keV and later by Watanbe et al. [1984] for synchrotron light of wavelength $\lambda = 30$ Å. A current enterprise is to produce x-ray beams as small as 0.1 μm using *"tapered glass capillaries"*. First successful experiments have been reported by Stern et al. [1988] and Thiel et al. [1989]. These focusing capillaries with a tapering angle of typically $\beta = 0.05$ mrad may produce intense x-ray beams with a tiny cross section that can be of considerable future importance in microdiffraction, x-ray microscopy or radiography. For an introduction into this exciting field I recommend the article by Stern et al. [1988].

2.2 Kinematic Theory of Grazing Angle Scattering

The demand of the surface science community for an "easy-to-apply" grazing incidence x-ray scattering theory was realized by Vineyard [1982] who revived the socalled "*Distorted Wave Born Approximation*" (DWBA). The simple idea behind this approach is to take the distorted wave (see Newton [1966]) as the "incoming beam" which is allowed to encounter a kinematic scattering cross section. Of course, this approach is only valid for crystals with a certain nonzero mosaicity, for perfect crystal surfaces the dynamical scattering theory which naturally includes the regime of total external reflection has to be applied. However, since this work is mostly concerned with nonperfect crystals we will touch upon the dynamical theory only very briefly (see Sect. 3.4.1). While Vineyard envisaged in his treatment only one grazing angle (α_i), the DWBA was completed to an arbitrary scattering geometry by Dietrich and Wagner [1983, 1984] and to the case of rough surfaces by Sinha et al. [1988]. (Note in this context also the elaborate theoretical study by Maradudin and Mills [1975] on the scattering of electromagnetic waves from the rough surface of a semi-infinite medium.) In the following I will sketch the essence of the derivation of the general scattering cross section in DWBA and will give some solutions of it for those scattering geometries which are relevant in the experiment.

2.2.1 Distorted Wave Born Approximation (DWBA)

Consider Fig. 2.5 which depicts the case of simultaneous total reflection and diffraction of x-rays: As before the incident beam $E_i(r)$ hits the crystal surface under a shallow angle α_i below the critical angle and creates by this a totally reflected beam $E_r(r)$ and the distorted wave $E_t(r')$ which travels according to (2.12) parallel to the surface within a thin skin at and underneath the surface. Whenever the distorted wave encounters a near-surface scattering cross section, the associated scattering intensity would be observable under a scattering angle 2θ parallel to the surface and another grazing (exit) angle α_f. Note that we introduce now the so far excluded part $\Delta n(r')$ of the x-ray index of refraction which is responsible for the scattering phenomena. Since the DWBA describes the kinematic scattering of the distorted wave within the first Born approximation, it is consequently governed by the following inhomogeneous Helmholtz equation

$$\text{rot rot } E_f(r) + k^2 n^2 E_f(r) = -k^2 \Delta n^2(r') E_t(r') \quad . \tag{2.20}$$

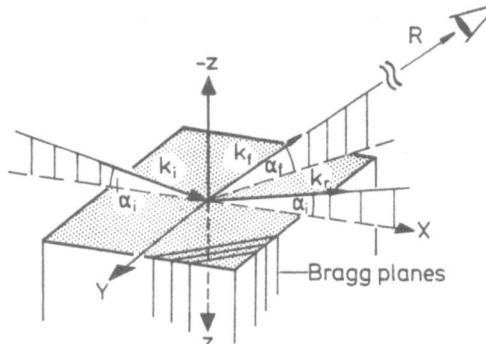

Fig. 2.5. Glancing angle scattering geometry: k_i, k_r and k_f are the wavevectors of the incident, reflected and scattered waves, respectively

It is most important to realize that (2.20) takes all refraction effects at the surface exactly into account, while it treats the scattering by single scattering events.

In the previous section we saw how the incident beam experiences the transmissivity of the interface between the vacuum and the matter. By applying again a reciprocity argument and reversing time we must follow that the same must be true for the scattered beam whenever passing through the surface. This type of argument goes back to Helmholtz' reciprocity principle in optics [Helmholtz v., 1886]:

> "If the source and the point of observation are interchanged, the same amplitude would result."

We expect thus that the scattered wave $E_f(R)$ should be symmetric with respect to the interchange of the indices i and f. As one knows from classical electrodynamics, the far field of a radiating dipole at an arbitrary distant point of observation $(R \to \infty)$ is $E_f(R) \propto (s \times p_i) \times s$, where $s = R/R$ and p_i is the polarization vector of the incident radiation. In our case, $p_i = \mathcal{I}_i e_i$ $(e_i \equiv E_i/E_i)$, and the scattered wave experiences the transmission function of the surface, thus, $E_f(R) \propto \mathcal{I}_f(s \times \mathcal{I}_i e_i) \times s$. A rigorous calculation gives [Dietrich and Wagner, 1983],

$$E_f(R) = r_e G(R)(1 - s \cap s) \cdot \mathcal{I}_f \cdot \mathcal{I}_i \cdot E_i \int_{z \geq 0} \Delta \varrho_e(r') e^{iQ' \cdot r'} d^3 r' \qquad (2.21)$$

with $Q' \equiv k_f' - k_i'$ denoting the scattering vector in the crystal and

$$k_f' = k \begin{pmatrix} \cos \alpha_f \cos 2\theta \\ \cos \alpha_f \sin 2\theta \\ -sinr \, \alpha_f \end{pmatrix} \qquad (2.22)$$

the refracted wave vector of the scattered wave $\left(\Delta \varrho_e(r') \right.$ is the spatial variation of the electron density). The propagator $G(R) = -\exp(ik \cdot R)/R$ represents as usual the amplitude observed at R due to a point of unit scattering strength in the crystal:

$$\text{rot rot } G(r) + k^2 G(r) = 4\pi \, \delta(r - R) \qquad \text{in the halfspace } z < 0.$$

In the coordinate system of Fig. 2.5 the transmission tensor $\mathcal{I}_f$ has now, because of the nonzero inplane scattering angle 2θ, a nondiagonal form

$$\mathcal{I}_f = \begin{pmatrix} T_{fa} & T_{fc} & 0 \\ T_{fc} & T_{fb} & 0 \\ 0 & 0 & T_{fz} \end{pmatrix} \tag{2.23}$$

with the elements

$$T_{fa} = T_{fx} \cos^2 2\theta + T_{fy} \sin^2 2\theta$$

$$T_{fb} = T_{fx} \sin^2 2\theta + T_{fy} \cos^2 2\theta \tag{2.24}$$

$$T_{fc} = \left(T_{fx} - T_{fz}\right) \sin 2\theta \cos 2\theta$$

and T_{fx}, T_{fy} and T_{fz} obtained from (2.11) after replacing the index i by f.

The standard polarization tensor $(1 - s \cap s)$ is as usual composed out of the unit tensor 1 and the self-dyadic product of s (see e.g. [Schiff, 1968]). Note that (2.21) degenerates to the bulk form for α_i and α_f much larger than α_c, then $\mathcal{I}_f \cdot \mathcal{I}_i \to 1$ and $Q' \to Q$ (the integral then extends over the full space).

The expression (2.21) for the scattered wave E_f allows a straightforward interpretation of scattering intensities observed under the condition of total external reflection, in particular because the applied first Born approximation retains the kinematic structure amplitude

$$F(Q') \equiv \int_{z \geq 0} \Delta\varrho_e(r') e^{iQ' \cdot r'} d^3 r' \tag{2.25}$$

with the only modification that Q' is used instead of the vacuum scattering vector Q as in conventional kinematic x-ray scattering. The measured intensity follows from (2.21) after performing a Kirchhoff integral:

$$I(Q') = I_0 r_e^2 S(Q') |t_{fi}|^2 \tag{2.26}$$

with the polarization-transmission factor

$$t_{fi} = (1 - s \cap s) \cdot \mathcal{I}_f \cdot \mathcal{I}_i \cdot e_i \quad , \tag{2.27}$$

$I_0 = E_i^2$ (photons/s·mm^2) and $S = \langle F^* F \rangle$ the kinematic struture factor and the brackets $\langle \dots \rangle$ denoting the thermal average. From (2.26) the kinematic scattering intensity can be deduced for any given grazing angle scattering geometry. We will give here the results for the case of two grazing angles α_i and α_f close to the critical angle α_c, thus $\alpha_{i,f} = O(10^{-3})$. Accordingly, we will evaluate E_f only up to linear order in $\alpha_{i,f}$.

For an incident x-ray wave polarized perpendicular to the plane of incidence, $e_{i\perp} = (0, 1, 0)$, we find

$$E_{f\perp} = r_e G(R) F(Q') T_{iy} \cdot \begin{pmatrix} -T_{fy} \sin 2\theta \cos 2\theta \\ T_{fy} \cos 2\theta \cos 2\theta \\ T_{fx} \sin \alpha_f \sin 2\theta \end{pmatrix} \tag{2.28}$$

and

$$|t_{\hat{\mathrm{f}}}^{\perp}|^2 = |T_{iy}|^2 \left(|T_{fy}|^2 \cos^2 2\theta + |T_{fx}|^2 \sin^2 \alpha_{\mathrm{f}} \sin^2 2\theta \right) \quad . \tag{2.29}$$

The second term in (2.29) is of order $O(\alpha_{\mathrm{c}}^2)$. When we neglect it, (2.29) yields a particularly simple form of the scattered intensity:

$$I_{\perp}(\boldsymbol{Q}') = I_0 \, r_{\mathrm{e}}^2 |T_{fy}|^2 \, |T_{iy}|^2 \, S(\boldsymbol{Q}') \cos^2 2\theta \tag{2.30}$$

with the familiar polarization factor $\cos^2 2\theta$ which renders $\boldsymbol{I}_{\perp} = 0$ at $2\theta = 90°$.

The expressions become a little more complicated for incident x-rays polarized parallel to the plane of incidence, $e_{i\parallel} = (\sin \alpha_{i}, 0, \cos \alpha_{i})$:

$$\boldsymbol{E}_{\mathrm{f}\parallel} = r_{\mathrm{e}} G(R) F(\boldsymbol{Q}')$$
$$\times \begin{pmatrix} T_{ix} T_{fy} \sin \alpha_{i} \sin^2 2\theta - T_{iz} T_{fz} \sin \alpha_{\mathrm{f}} \cos 2\theta \\ -T_{ix} T_{fy} \sin \alpha_{i} \sin 2\theta \cos 2\theta - T_{iz} T_{fz} \sin \alpha_{\mathrm{f}} \sin 2\theta \\ T_{ix} T_{fx} \sin \alpha_{i} \sin \alpha_{\mathrm{f}} \cos 2\theta - T_{iz} T_{fz} \end{pmatrix} \tag{2.31}$$

and thus

$$|t_{\hat{\mathrm{f}}}^{\parallel}|^2 = |T_{iz}|^2 |T_{fz}|^2 + \left(|T_{ix}|^2 |T_{fy}|^2 \sin^2 \alpha_{i} \sin^2 2\theta \right. \tag{2.32}$$
$$\left. + |T_{iz}|^2 |T_{fz}|^2 \sin^2 \alpha_{\mathrm{f}} - 2| T_{ix} T_{fx} T_{iz} T_{fz} | \sin \alpha_{i} \sin \alpha_{\mathrm{f}} \cos 2\theta \right) \quad .$$

However, in our case, the longish term in brackets $(\dots)$ is again of order $O(\alpha_{\mathrm{c}}^2)$ and negligible leaving behind again a simple expression

$$I_{\perp}(\boldsymbol{Q}') = I_0 r_{\mathrm{e}}^2 |T_{iz}|^2 |T_{fz}|^2 S(\boldsymbol{Q}') \tag{2.33}$$

for the scattering intensity with, of course, no polarization factor.

2.2.2 Depth Profiles of Surface Scattering

The interesting quantity is the scattering amplitude $F(\boldsymbol{Q}')$ which provides the information on surface-related microscopic correlations. By dividing the scattering vector $\boldsymbol{Q}'$ into the inplane component $\boldsymbol{Q}_{\parallel}$ and normal component Q'_z we rewrite (2.25) as

$$F(\boldsymbol{Q}_{\parallel}, Q'_z) = \int_{-\infty}^{\infty} \int_{z \geq 0} \Delta\varrho_{\mathrm{e}}(\boldsymbol{r}_{\parallel}, z) \, \mathrm{e}^{-i\boldsymbol{Q}_{\parallel} \cdot \boldsymbol{r}_{\parallel}} \, d^2 \boldsymbol{r}_{\parallel} \, \mathrm{e}^{-iQ'_z z} dz$$
$$\equiv \int_{z \geq 0} F(\boldsymbol{Q}_{\parallel}, z) \, \mathrm{e}^{-i\kappa'_z z} \mathrm{e}^{-z/\Lambda} dz \tag{2.34}$$

with $\kappa'_z \equiv \mathrm{Re}\{Q'_z\}$ and

$$\Lambda \equiv |\mathrm{Im}\{Q'_z\}|^{-1} \quad , \tag{2.35}$$

which determines the depth, where the observed scattering originates from. In order to distinguish this length from the penetration depth of the evanescent wave (which depends only on the incidence angle α_i) we call Λ the "scattering depth". According to (2.34) grazing angle scattering experiments provide the complex Laplace transform of the scattering law $F(Q_\parallel, z)$ within a surface region given by the scattering depth Λ. The properties of Λ have been discussed in literature in detail [Dosch et al., 1986; Dosch, 1987]: Again, as a consequence of reciprocity, both grazing angles $\alpha_{i,f}$ enter symmetrically the quantity Λ,

$$\Lambda = \frac{\lambda}{2\pi(l_i + l_f)} \tag{2.36}$$

with $l_{i,f}$ from (2.13) for both indices i and f. Figure 2.6a shows the behaviour of Λ as a function of α_f for different values of β/δ and α_i: While it is intuitively clear that Λ is small for $\alpha_i < \alpha_c$ because of the evanescent wave effect, one finds that the observation of the radiation at grazing exit angles $\alpha_f < \alpha_c$ has the same effect, in other words, radiation from bulk atoms cannot penetrate into the vacuum, if the exit angle is below the critical angle of the interface. This interesting phenomenon has implicitly been adressed already by Sommerfeld [1972] in discussing the "radio problem" (a source in the vicinity of a plane interface). Figure 2.6a demonstrates that a whole range of depths can be profiled by a proper experimental grazing angle setup. It was pointed out recently that the most efficient way of obtaining such depth profiles is by using a position sensitive detector which measures the inplane scattering (at a fixed scattering angle $Q_\parallel$) in an α_f-range between 0 and typically $4\alpha_c$ [Dosch et al., 1986; Dosch, 1987].

We discuss the principle shortly for two opposite types of scattering laws, kinematic Bragg scattering with a very pronounced Q-dependence and incoherent (as Compton- or fluorescence) scattering with a negligible Q-dependence:

Consider now Bragg planes perpendicular to the surface under investigation (Fig. 2.5). The ideal semi-infinite kinematic sum for Bragg scattering is simply evaluated,

$$\begin{aligned}
S_B(Q_\parallel, Q'_z) &= \sum_{m,n=0}^{\infty}\sum^{\infty} F_m F_n^* e^{iQ_\parallel \cdot (r_m - r_n)} e^{i(Q'_z z_m - Q'^*_z z_n)} \\
&= |F_{hkl}|^2 \delta(Q_\parallel - G_{hkl}) \frac{1}{|1 - e^{iQ'_z a_\perp}|^2} ,
\end{aligned} \tag{2.37}$$

with $a_\perp$ as the lattice constant normal to the surface, F_{hkl} the structure factor of the unit cell, G_{hkl} the associated reciprocal lattice vector and δ the Kronecker symbol. In the derivation of (2.37) we tacitly assumed that the Bragg condition parallel to the surface is not violated by moving along the Q_z-direction: Whenever the projections of k_i and k_f are equal (thus for $\alpha_i = \alpha_f \equiv \alpha$) the Bragg condition can be fulfilled exactly with the Bragg angle $\theta(\alpha) = \sin^{-1}(|G_{hkl}|/(2k \cos\alpha))$. On the other hand, for $\alpha_i \neq \alpha_f$ one gets approximately [Afanase'v and Melkonyan, 1983] $\sin 2\theta \Delta\theta \approx (\alpha_f^2 - \alpha_i^2)/2$, thus, in our scattering geometry, where $(\alpha_f^2 - \alpha_i^2) = O(10^{-6})$, the detuning $\Delta\theta$ from the exact parallel Bragg condition is typically of the order of some seconds of

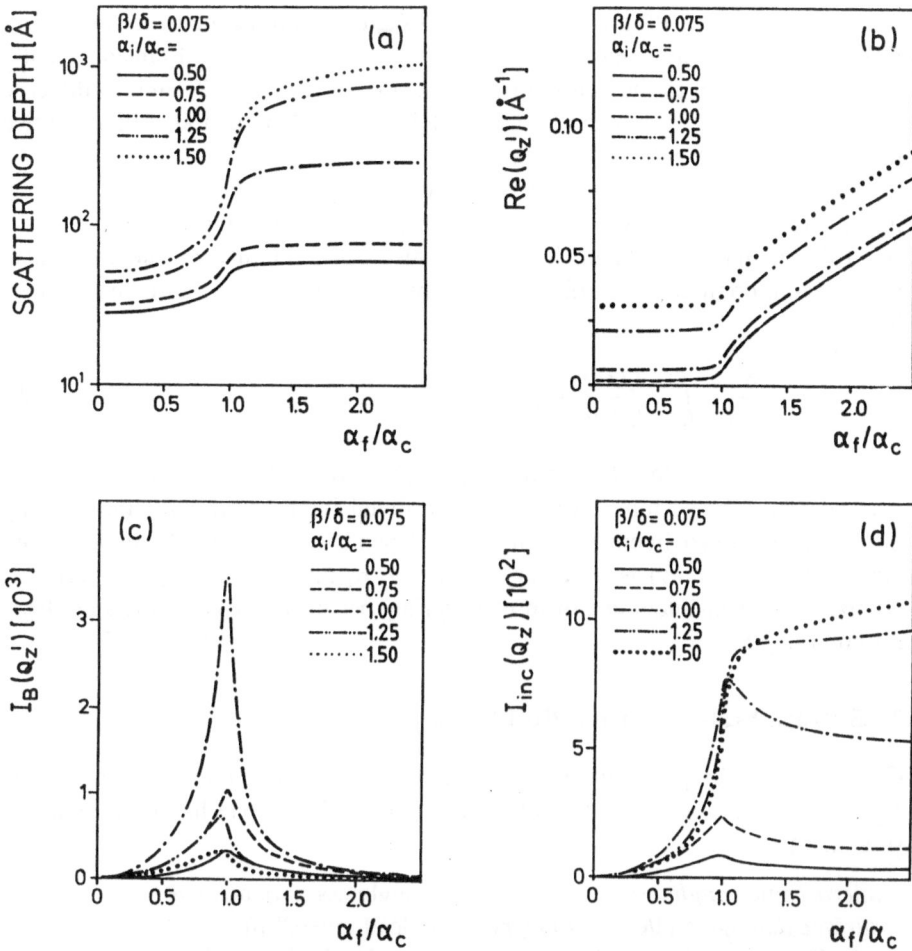

Fig. 2.6a-d. DWBA-scattering quantities as a function of α_f/α_c for different α_i/α_c: (a) Scattering depth (2.36); (b) real part of momentum transfer normal to the interface; (c) evanescent Bragg scattering law (2.37); (d) evanescent incoherent scattering law (2.38)

arc or some μrad which is many orders of magnitude smaller than the mosaicity of the crystal and can safely be neglected. In the case of a perfect crystal the Darwin width is comparable to $\Delta\theta$ and has to be taken into account.

For totally incoherent scattering we get instead [Dosch, 1987]

$$S_{\text{inc}}(Q'_z) = \sum_{m=0}^{\infty} \left| f_{\text{inc}} e^{-iQ'_z z_m} \right|^2 = \frac{f_{\text{inc}}^2}{1 - e^{-2a_\perp/\Lambda}} \propto f_{\text{inc}}^2 \Lambda \qquad (2.38)$$

(see also the discussion of (5.25) in Sect. 5.1.3). While S_{inc} does not depend on the real part of the scattering vector, it should be noted that S_{inc} has a distinct α_f-dependence which is implicitly given in (2.38) by the α_f-dependence of the scattering depth Λ. The α_f-profiles of the Bragg intensities (2.37) and the incoherent intensities (2.38) are shown in Fig. 2.6c,d for various incidence angles α_i. Both intensities exhibit an enhancement at $\alpha_f = \alpha_c$ ("Vineyard enhancement") which is attributed to the action of the transmission function T_f and, in the case of Bragg scattering, partly to the scattering law itself [Dosch et al., 1986]. Notice that $\kappa_z \equiv \mathrm{Re}\{Q'_z\} \simeq 0$ for $\alpha_{i,f} < \alpha_c$ (see Fig. 2.6b).

From the discussion above it should be clear that such α_f-resolved scattering intensities can be converted into depth profiles depending on the setting of (α_i, α_f). On the other hand, it becomes also apparent that α_f-integrated data have a more complicated surface sensitivity associated with an effective scattering depth

$$\Lambda^i_{\mathrm{eff}} = \frac{\lambda}{2\pi\, I_{\mathrm{int}}} \int_{\sigma_D} \frac{I_i(\alpha_f)}{l_i + l_f} \, d\alpha_f \qquad (2.39)$$

depending on the effective aperture $\sigma_D = \Delta z_D / R$ of the detector (Δz_D is the vertical detector aperture, $I_i(\alpha_f)$ the α_f-resolved intensity and I_{int} the actually observed integrated intensity). Since Λ^i_{eff} depends on the underlying scattering law, particular caution is necessary when e.g. α_f-integrated Bragg scattering is compared with α_f-integrated diffuse scattering, because they generally stem from different depths.

2.2.3 Bragg Scattering from Real Surfaces

Grazing angle Bragg scattering is sensitive to the structural properties of the crystal surface. When a real solid surface is considered, one has to account for (Fig. 2.7a):

– *microscopic roughness:*   (roughness parameter ϱ)
– *surface damage* and/or *oxid layers:*   ("dead layers" p)
– *misalignment of Bragg planes:*   (angular deviation $\Delta\phi$)
– *macroscopic waviness:*   (variation of surface normal Δn_s)

The *roughness effect* on $S_B(Q'_z)$ can be deduced by inspection of Fig. 2.7b which shows a schematic blow-up of a solid surface: Since the depth z_n varies with the lateral coordinate $r_\parallel$, the semi-infinite sum in (2.37) has to be written as the lateral average

$$S^\varrho_B(Q'_z) \propto \left\langle \sum_{m,n=0}^{\infty}\sum e^{-iQ'_z\left(a_\perp n + \Delta z(r_{1\parallel})\right)} e^{iQ'^*_z\left(a_\perp m + \Delta z(r_{2\parallel})\right)} \right\rangle_\parallel \qquad (2.40)$$

which has a simple analytical solution for a Gaussian $\Delta z(r_\parallel)$ (2.15), namely (see B.1)

$$S^\varrho_B(Q'_z) = e^{-2M_\varrho}\, S_B(Q'_z) \qquad (2.41)$$

with

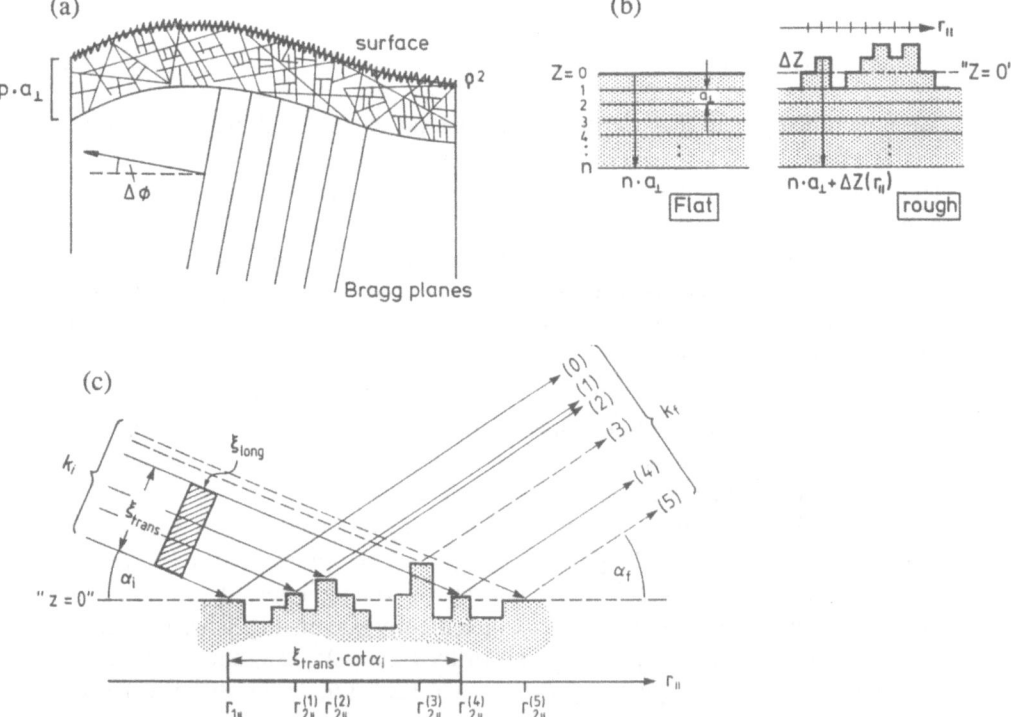

Fig. 2.7a-c. Schematic side view of a real surface: (a) General surface damage: "dead" surface layers p, surface waviness Δn_s, surface roughness ϱ, misalignment of Bragg planes $\Delta\phi$. (b) Solid-on-Solid model of a rough surface: $z = 0$ denotes the average surface (c) Glancing angle Bragg scattering from a rough surface and coherence effects: The Bragg scattered waves with wave vector k_f denoted (0), (1), (2) and (4) are excited from a surface area which is illuminated by a spatially coherent x-ray wave and give rise to the observation of a Debye–Waller factor $\exp(-2M_\varrho)$ (2.41). The waves (3) and (5) are incoherent to the others, since they stem from outside the coherence volume $\xi^2_{\text{trans}} \times \xi_{\text{long}}$ (hatched region) of the impinging radiation (see text). The length $\xi_{\text{trans}} \cot \alpha_i$ is the largest lateral distance on the surface which may be "seen" coherently

$$2M_\varrho = \left(\kappa_z'^2 - \frac{1}{\Lambda^2}\right)\varrho^2 \tag{2.42}$$

with the roughness parameter ϱ and $\kappa_z' = \text{Re}\{Q_z'\}$ as before. The surface roughness-induced diffuse scattering will be discussed in the context of evanescent neutron scattering (Sect. 3.3.1) and can easily translated by the reader into the x-ray case.

A more severe surface irregularity is a *strongly damaged surface region* which does not contribute to the Bragg intensity at all. If we attribute a thickness of $pa_\perp$ to these "dead layers" (as created e.g. by the mechanical polishing of a single crystal) we find the modified kinematic sum (see [Dosch, 1987])

$$S_B^p(Q_z') \propto \sum_{m,n=p}^{\infty}\sum^{\infty} e^{-iQ_z' a_\perp n} e^{iQ_z'^* a_\perp m} = e^{-2M_p} S_B(Q_z') \tag{2.43}$$

19

with

$$2M_p = 2pa_\perp / \Lambda \quad .$$ (2.44)

We note already here that wetting transitions (as surface melting transitions) are characterized by $p(t) = p_0 \ln |1/t|$ which then leads to $S_B^p(Q_z') = t^{2p_0 a_\perp/\Lambda} S_B(Q_z')$ (t is the reduced temperature). This powerlaw behaviour will be discussed in detail in Chap. 5.

Due to uncertainties in aligning the surface during the preparation of the crystal surface, one has to account for a certain *misorientation $\Delta\phi$ of the Bragg planes* with respect to the surface normal, by setting $Q_z' \rightarrow Q_z' - \Delta\phi |G_{hkl}| \equiv Q_{z\phi}'$ for small values of $\Delta\phi$.

So far we ignored the *waviness of a surface* as expressed by the variation Δn_s of the surface normal. The question, whether a surface structure is "seen" by the impinging x-ray wave as roughness or as waviness, can be answered by considering the coherence length of the radiation.

As one knows, only a pointlike and ideally monochromatic source emits light of infinite coherence length, accordingly, since any real x-ray source has a nonzero linear extension Δd_s and a nonzero wavelength spread $\Delta\lambda$, the coherence of the emitted radiation is finite. Conveniently one distinguishes between the longitudinal coherence length ξ_{long}, given by $\xi_{\text{long}} = \lambda^2/\Delta\lambda$, and the transverse coherence length ξ_{trans}, given by $\xi_{\text{trans}} = \lambda R_s/\Delta d_s$ (R_s is the distance between the x-ray source and the sample). Typical x-ray stations in modern synchrotron radiation laboratories are charcterized by $\Delta\lambda/\lambda \simeq 5 \times 10^{-4}$, $\Delta d_s \simeq 0.3$ mm and $R_s \simeq 30$ m (see Sect. 2.4.3), thus for x-ray wavelengths of $\lambda = 1$ Å $- 2$ Å we have to consider x-ray coherence lengths of $\xi_{\text{long}} \simeq 0.2$ μm $- 0.4$ μm and $\xi_{\text{trans}} = 10$ μm $- 20$ μm. Surface irregularities related with these length scales create a coherent scattering pattern and are therefore seen as "rough", whereas for larger surface structures the scattering intensity has to be averaged over the associated variations of the incidence and exit angles $\Delta\alpha_{i,f}(\xi_c)$,

$$I(\alpha_{i0}, \alpha_{f0}) = \iint I(\alpha_{i0} - \alpha_i, \alpha_{f0} - \alpha_f) \, w[\Delta\alpha_i(\xi_c)] \, w[\Delta\alpha_f(\xi_c)] \, d\alpha_i \, d\alpha_f$$

(2.45)

with $w(x)$ as proper weights. In other words, the borderline between roughness and waviness depends on the spectral purity $\Delta\lambda$ and on the *"parallaxis angle"* $\Delta d_s/R_s$ of the light we use. This is illustrated in Fig. 2.7c.

2.3 Some Applications of Grazing Angle Scattering

An interesting experiment to test the concept of *"microreversibility"* (see [Messiah, 1970]) in the presence of evanescent x-ray waves has been performed by Becker and coworkers [1983]. They could demonstrate that the fluorescence yield emitted under grazing exit is the same as the total absorption observed under graz-

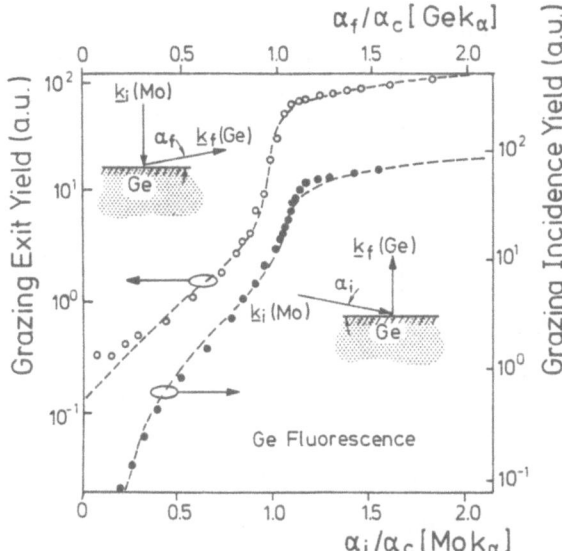

Fig. 2.8. Experimental test of the concept of microreversibility on a Ge single crystal surface [Becker et al., 1983]. For explanation see text

ing incidence. Figure 2.8 shows their results as obtained on a Ge single crystal surface. Since absorption and fluorescence are completely incoherent processes, one would expect that the observed intensities should be describable by the action of the transmission function and the scattering depths as discussed above. The theoretical curves (full lines in Fig. 2.8) have indeed been calculated according to (2.38) and confirm the experimental findings. It should be noted, however, that the original reciprocity principle holds strictly only for purely elastic scattering. When inelastic processes are present (as absorption in this case), the energy has to be rescaled upon applying time reversal.

Evanescent fluorescence has been used by Brunel [1986] and Brunel and de Bergevin [1986] to study impurity concentration profiles $C(z)$ in semiconductors with an experimental setup similar to that depicted in the inset of Fig. 2.8. The fluorescence intensity $I_F(\alpha_i)$, as measured with the SiLi solid state detector normal to the surface, provides directly the quantity

$$I_F(\alpha_i) \propto \int_0^\infty C(z)\, e^{-z/\ell_i}\, dz \quad , \tag{2.46}$$

thus, the Laplace transform of the concentration profile $C(z)$ with respect to the penetration depth of the evanescent x-ray wave (see also [Yun and Bloch, 1990]). The surface sensitivity of such experiments could certainly be enhanced in the future by exploiting scattering depth effects with two grazing angles [Sasaki and Hirokawa, 1990].

First depth profiles of near surface Bragg and diffuse synchrotron x-ray scattering have been reported by Dosch et al. [1986] and Dosch [1987]. They investigated by way of example the $\alpha_{i,f}$-dependence of the $(2\bar{2}2)$ Bragg scattering from a mirrorlike $Fe_3Al(110)$ single crystal surface and interpreted the measured

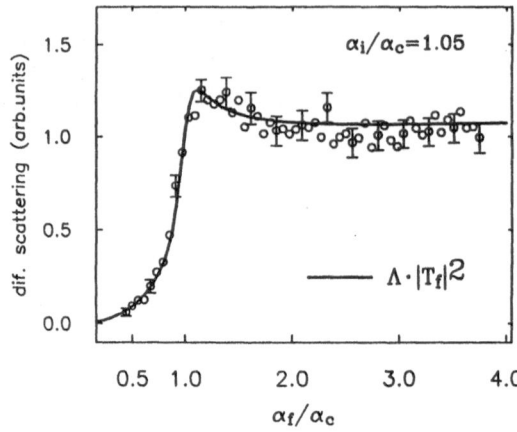

Fig. 2.9. Evanescent incoherent scattering observed at an $Fe_3Al(110)$ surface at grazing angles [Dosch, 1987]. The full line is calculated according to (2.38)

scattering intensities in terms of the DWBA equation (2.26). The authors introduced the surface disorder term $\exp(-2pa_\perp/\Lambda)$ to account for surface damage and oxid layers and pointed out the potential of obtaining detailed depth profiles of near-surface atomic correlations by α_i- and α_f-resolved grazing angle scattering. Figure 2.9 displays the observed α_f-dependence of the diffuse surface scattering far away from the Bragg condition which has, by reciprocity, the same form as the α_i-profile of the Ge-fluorescence shown in Fig. 2.8.

These kind of experiments have been followed up by Lied and coworkers [1989] who applied this scheme to characterize the final polishing steps in the mechanical surface preparation procedure of FeAl surfaces. Two exemplary α_f-profiles of the $(2\bar{2}2)$ Bragg reflection of a (110) surface after a $1\,\mu m$ Al_2O_3 polish and after a $50\,nm$ SiO_2 treatment are presented in Fig. 2.10 together with DWBA Bragg profiles. The results of the curve fitting procedure reveal for example the thickness $(pa_\perp)$ of the remaining surface damage, the roughness (ϱ) and waviness (Δn_s) of the surface and the misorientation $(\Delta\phi)$ of the Bragg planes with respect to the surface. It is evident that such a depth selective monitoring of the

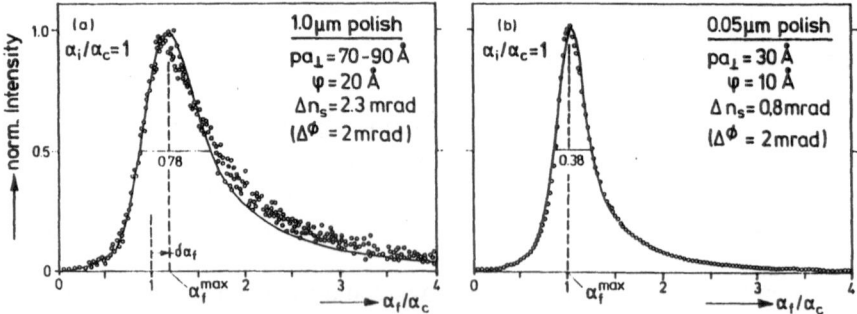

Fig. 2.10. Grazing angle Bragg scattering from an $Fe_3Al(110)$ surface as a monitor of the surface damage induced by the mechanical polish (a) with $1.0\,\mu m$ and (b) with $0.05\,\mu m$ Al_2O_3 [Lied et al., 1989]. The full lines are theoretical curves based on (2.40–45)

various polishing steps, which is unrivaled by other surface techniques, would be very desirable in surface preparation laboratories. A GID surface characterization station of this kind may one day be implemented in crystal preparation laboratories located at future dedicated synchrotron sources.

Many other interesting applications of evanescent x-ray diffraction have emerged in the last years. I mention here the study of ion-implanted Si crystals which has recently been reported by Grotehans et al. [1989] who measured depth profiles of near-surface Huang diffuse scattering evoked by the lattice distortions around the irradiation defects, the study of x-ray transmissivity under grazing incidence [Rieutord, 1990] and the study of small angle x-ray scattering from surfaces and thin films [Levine et al., 1989]. Incidently it should be noted that evanescent diffuse x-ray scattering has already been reported in the early 60'ies by Yoneda [1963] who observed small angle scattering at grazing exit angles from various solid surfaces. Due to the action of the transmission function the tiny diffuse intensity (presumably caused by surface roughness) was peaked at the critical angle of the substance. While Yoneda mystified his observation somewhat (*"anomalous surface reflection"*), Warren and Clarke [1965] and Guentert [1965] provided essentially the correct qualitative explanation (see Sect. 3.3.1). For a recent discussion see also Gorodnichev et al. [1988].

2.4 Experimental Considerations

2.4.1 Glancing Angle Scattering Geometry

The prominent features of a surface sensitive x-ray scattering experiment are the two glancing angles which are close to the critical angle of the matter. We already mentioned that typically $\alpha_c = O(10^{-3})$ for common wavelengths. Many of the experimental challenges arise from the fact that the required angle of incidence has to be as small as some fractions of a degree with an accuracy of, say, $\Delta\alpha_i/\alpha_c \leq 0.05$. Figure 2.11 shows a side view of the scattering geometry at the sample position: The intercepted beam height is $L_s \cdot \alpha_c = O(\mu m)$ for a typical sample surface diameter of $L_s = 10$ mm, therefore, a slit (σ_{iv}) in

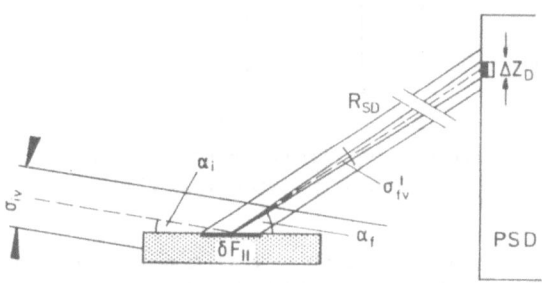

Fig. 2.11. Side view of the grazing angle scattering geometry. $\delta F_\parallel$ is the illuminated surface area (see Fig. 2.12)

front of the sample taylors the vertical beam dimension to a proper size[2], say $\sigma_{iV} = 20 - 45$ μm. In addition, the vertical beam divergences σ'_{iV}, σ'_{fV} have to be extremely small ($\simeq \alpha_c/20$), since they determine Q'_z and thus the surface sensitivity. σ'_{fV} is given by the spatial resolution Δz_D of the position sensitive detector (PSD) which is at a distance R_{SD} away from the sample, i.e., $\sigma'_{fV} = \Delta z_D/R_{SD}$. PSD systems based on carbon-coated glass wires or on metal wires achieve $\Delta z_D^g \cong 30 - 45$ μm at a total sensitive length of $\ell_D = 40$ mm, thus, provide relative spatial resolutions around $\delta s \equiv \Delta z_D^g/\ell_D \simeq (0.5 - 1.0) \times 10^{-3}$ (which is the relevant figure of merit). For a required α_f-range of $r\alpha_c$ (typically $r = 4$) the minimal attainable α_f resolution is then $\Delta \alpha_f/\alpha_c = r\delta s$ independent of the used wavelength and the sample properties. The illuminated surface area $\delta F_{\parallel}$ is shown in Fig. 2.12: The two horizontal slits σ_{iH} and σ_{fH} assure that only a spot in the middle of the sample surface is seen by the detector and the sample edges are avoided. Then $\delta F_{\parallel}$ is simply given by

$$\delta F_{\parallel} = \sigma_{iH}\sigma_{fH}/\sin 2\theta \quad . \tag{2.47}$$

For a given set of $\sigma_{i,fH}$ this concept works, however, only as long as the scattering angle 2θ is larger than a critical angle $2\theta^*$, where $\delta F_{\parallel}$ just touches the sample borders (Fig. 2.12b). If $2\theta < 2\theta^*$, $\delta F_{\parallel}$ becomes complicated (sample shape dependent) and unwanted scattering from the sample edges (bulk scattering) may be picked up by the detector. The useful approximate relation

$$2\theta^*[\text{rad}] \simeq \frac{\sigma_{iH} + \sigma_{fH}}{L_s} \tag{2.48}$$

permits for a given sample surface with a typical dimension L_s the appropriate adjustment of the horizontal slits in order to obtain a well-defined surface scat-

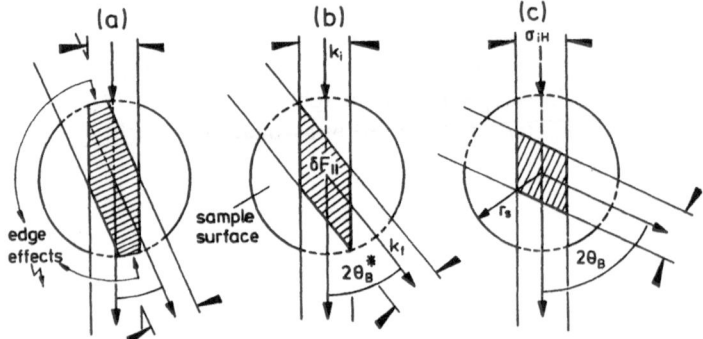

Fig. 2.12a-c. Top view of the grazing angle scattering geometry and the effect of the horizontal slits. (a)–(c) show the active surface area $\delta F_{\parallel}$ for increasing inplane scattering angle 2θ

---

[2] In discussing the properties of synchrotron radiation the use of σ-values, as done in this section, is common practice. If the reader is more accustomed to FWHM-values, he has to multiply all σ-values by the factor $2\sqrt{(2 \ln 2)} = 2.3548$.

tering signal down to a scattering angle $2\theta^*$. (E.g., for $L_s = 10$ mm and a required lower value $2\theta^* \simeq 10°$ one finds $\sigma_{\mathrm{iH}}, \sigma_{\mathrm{fH}} \simeq 1$ mm).

In the surface plane we request a high resolution in $Q_\|$ for precise measurements of in-plane structure factors, like e.g., Bragg profiles or the lineshape of surface critical diffuse scattering. Differentiating Bragg's equation we obtain

$$\Delta Q_\| / Q_\| = \Delta\lambda/\lambda + \left(\sigma'_{\mathrm{iH}} + \sigma'_{\mathrm{fH}}\right) \frac{\cot\theta}{2} \quad , \qquad (2.49)$$

which asks as in bulk scattering experiments for a highly monochromatic $(\Delta\lambda/\lambda)$ and horizontally collimated (σ'_{iH}) beam with typically $\Delta\lambda/\lambda \simeq 10^{-4} - 10^{-3}$ and $\sigma'_{\mathrm{iH}} \leq 1$ mrad (σ'_{fH} is given by the detector slit). Well, when we summarize this, we call for a monochromatic x-ray beam with a cross section of $\sigma_{\mathrm{iV}}\sigma_{\mathrm{iH}} \simeq 50\,\mu\mathrm{m}\cdot 1\mathrm{mm}$ and a divergence of $\sigma'_{\mathrm{iV}}\sigma'_{\mathrm{iH}} \simeq 0.3\,\mathrm{mrad}\cdot 0.5\,\mathrm{mrad}$ and which contains still enough photons/s to allow the observation of the scattering from a surface layer of thickness $\Lambda \simeq 50\,\text{Å}$. What we ask for is in other words that the spectral photon-phase space $\varepsilon_\lambda \equiv \sigma_{\mathrm{iV}}\sigma_{\mathrm{iH}}\sigma'_{\mathrm{iV}}\sigma'_{\mathrm{iH}}\Delta\lambda/\lambda$ ("*spectral emittance*") is as small, or the "*spectral brightness*",

$$B_\lambda \equiv \frac{N/t}{\varepsilon_\lambda} \quad (N/t = \text{photons/s}) \quad , \qquad (2.50)$$

of the x-ray source is as large as possible. The experts in the field know of course that we introduced with B_λ the trade-off quality of modern synchrotron sources [Koch, 1983]. This is best demonstrated in Fig.2.13, where the spectral brightness of various synchrotron sources are compared, the Hamburg Synchrotron Radiation Laboratory (HASYLAB) at the Deutsches Elektronen Synchrotron (DESY)

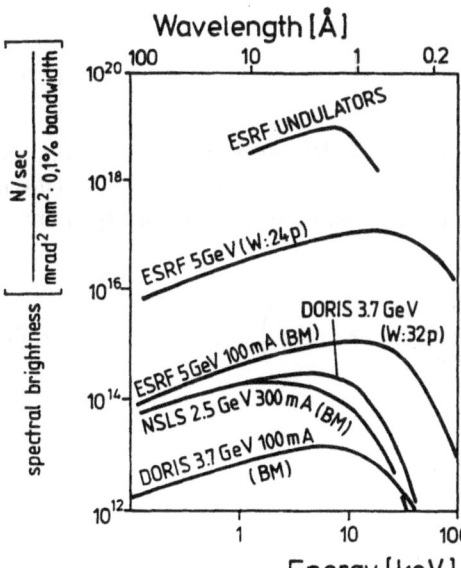

Fig. 2.13.
Spectral brightness of some synchrotron radiation sources (BM = bending magnet, W: 32p = 32-pole wiggler): storage ring DORIS (HASYLAB, Hamburg); National Synchrotron Light Source (NSLS, Brookhaven, NY): European Synchrotron Radiation Facility (ESRF, Grenoble, France)

in Hamburg, the National Synchrotron Light Source (NSLS) at Brookhaven (New York) and the European Synchrotron Radiation Facility (ESRF) at Grenoble which will become operational in 1994. If we plotted the brightness of the CuK_α characteristic line of a conventional x-ray generator (rotating anode with micro-focusing) we would end up at $I_0 \cong 10^{10}$ photons/(s · mm^2 mrad^2 0.1%) which is out of the lower scale of Fig. 2.13. The small beam divergence vertical to the electron orbit is given by the kinetic energy of the orbiting electron (typically some GeV) scaled to its rest mass,

$$\sigma'_V(1\,e^-) \cong \gamma^{-1} \equiv \frac{m_0 c^2}{E_{\text{kin}}} \cong \frac{511\,\text{keV}}{3\,\text{GeV}} \cong 0.2\,\text{mrad} \tag{2.51}$$

(m_0 is the electron rest mass).

2.4.2 Integrated Grazing Angle Bragg Intensity

A reliable measurement and interpretation of integrated Bragg intensities observed under total external reflection condition is particularly important, when near-surface long range order, or, in the case of phase transitions, near-surface order parameter profiles are considered. As in the bulk case [Guinier, 1963] we perform an ω-scan (rotation of the sample around the reciprocal lattice vector with a fixed detector at the 2θ position): An incident beam with the properties described before impinges at a grazing angle α_i onto the surface of a single crystal which is properly aligned for Bragg reflection; the Bragg scattering is detected by a PSD with a pixel size $A_D = \Delta z_D \cdot W_\parallel$ (where $W_\parallel$ is assumed to be large enough to collect all contributions to the Bragg intensity). While we are rotating the crystal around its surface normal with a constant angular velocity $\varphi = d\omega/dt$, we record the evanescent intensity scattered into the detector, thus we measure

$$P_{\text{GID}}^i = \int_{\text{Peak}} |E_f|^2 \, dt \int_{A_D} d^2 A \tag{2.52}$$

(E_f is the scattered amplitude (2.21)). Since $dA = R_{\text{SD}}^2 \, d\xi \, d\gamma$ and $dt = d\omega/\varphi$, we can replace the integration segment $dt \, d^2 A$ by $(R_{\text{SD}}^2/\varphi) \, d\gamma \, d\omega \, d\xi \, (d\gamma = d\alpha_i + d\alpha_f)$. The angular variations $d\omega$ and $d\xi$ detune the parallel scattering vector $Q_\parallel$ in linear order by

$$dQ_\parallel = \begin{pmatrix} dQ_x \\ dQ_y \\ 0 \end{pmatrix} = k \begin{pmatrix} -\sin 2\theta \, \cos \alpha_f \, d\xi \\ \cos 2\theta \, \cos \alpha_f \, d\xi - \cos \alpha_i \, d\omega \\ 0 \end{pmatrix} , \tag{2.53}$$

therefore,

$$d\omega \, d\xi = \frac{dQ_x \, dQ_y}{k^2 \sin 2\theta \, \cos \alpha_i \, \cos \alpha_f} \tag{2.54}$$

and

$$P_{\text{GID}}^{i} = \frac{I_0 P r_e^2}{\varphi k^2 \sin 2\theta} |F_{hkl}|^2 \int\int_{-\infty}^{\infty} dQ_x dQ_y \, \delta\left(Q_{\parallel} - G_{hkl}\right)$$

$$\times \int_0^{r\alpha_c} d\alpha_f \left(\frac{|T_i|^2 |T_f|^2}{\cos \alpha_i \cos \alpha_f |1 - e^{iQ_z' a_\perp}|^2}\right) \tag{2.55}$$

with the polarization factor $P = (1; \cos^2 2\theta)$. Notice that we extended the finite integration over the Bragg peak now to $\pm\infty$ (which gives the same result, since away from the Bragg peak we don't assume any scattering) and obtain for it $(2\pi/a_1)(2\pi/a_2)N_1 N_2 = 4\pi^2 N_1 N_2/F_\Omega$ with $F_\Omega = a_1 a_2$ as the unit mesh in the surface plane and N_1, N_2 the number of surface atoms illuminated in the surface area $\delta F_{\parallel}$, thus, $N_1 N_2 F_\Omega = \delta F_{\parallel}$. Inserting $\delta F_{\parallel}$ (2.47) in (2.55) we get at an incidence angle α_i a total number of photons,

$$P_{\text{GID}}^{i} = \frac{I_0 \lambda^2 P r_e^2}{F_\Omega^2 \varphi \sin^2 2\theta} |F_{hkl}|^2 \sigma_{iH} \sigma_{fH} \alpha_c$$

$$\times \int_0^r d(\alpha_f/\alpha_c) \left(\frac{|T_i|^2 |T_f|^2}{\cos \alpha_i \cos \alpha_f |1 - e^{iQ_z' a_\perp}|^2}\right) \quad , \tag{2.56}$$

collected in the PSD covering an α_f-range between 0 and $r\alpha_c$. The dimensionless integral in (2.56) has to be "shot" (evaluated numerically). We give an example which scattering intensity can be expected:

We consider the Al(002) Bragg reflection of a (110) single crystal surface measured at a Wiggler beam line with a focused monochromatic beam ($\lambda = 1.541$ Å). If we take $\sigma_{iH}, \sigma_{fH} \simeq 1$ mm and assume a photon flux of $I_0 = 1 \cdot 10^9$ photons/(s mm^2) on the sample surface, then we should measure at an incidence angle of $\alpha_i/\alpha_c = 0.5$ and a scan velocity of $\varphi = 0.1°$/s totally $P_{\text{GID}} \simeq 6 \cdot 10^3$ counts in the position sensitive detector. Thus, the observation of evanescent *Bragg* scattering usually poses no intensity problem and is also feasible with conventional x-ray sources (i.e. rotating anode with fine focus). As a matter of fact the grazing angle Bragg intensities shown in Fig. 2.10 have been measured using a 60 kW rotating x-ray generator. From (2.56) one also finds

$$P_{\text{GID}}^{i} \propto \left\{ \begin{array}{c} \cot^2 2\theta \\ \sin^{-2} 2\theta \end{array} \right\} \quad \text{for} \quad E_i = \left\{ \begin{array}{c} E_{i\perp} \\ E_{i\parallel} \end{array} \right\} \tag{2.57}$$

for the inplane 2θ-dependence of evanescent integrated Bragg scattering.

2.4.3 Experimental Station

Today surface x-ray scattering experiments are commonly performed with synchrotron radiation. Dedicated synchrotron laboratories are furnished with socalled "insertion devices", special magnetic structures called "wigglers" and "undulators" which produce taylored synchrotron x-radiation. The brilliant white x-ray beam usually passes through a sophisticated beam optics system, where it gets monochromatized by Bragg scattering from a single crystal arrangement and fo-

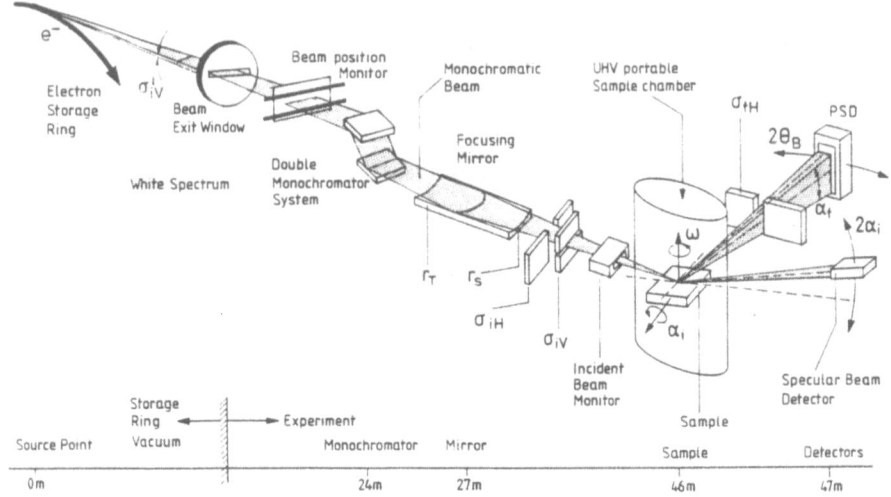

Fig. 2.14. Schematic view of an experimental station dedicated to surface x-ray diffraction studies (after Dosch et al. [1991a]); for explanation see text

cused into the experimental hutch by a total reflection mirror. The experimental hutch is furnished with a heavy weight diffractometer dedicated to surface work. Though all the individual components of such a beam line and their arrangement along the beam may vary between equivalent experimental stations depending on the local philosophy, the essentials are as in Fig. 2.14 which gives a schematic view of a GID experiment similar to that at the W1-beam line at HASYLAB which will be discussed now in a little more detail.

The white beam from a 32-pole wiggler leaves the storage ring vacuum and enters the beam optics section through an exit window. The monochromator system is composed out of two monochromator crystals in the socalled nondispersive arrangement which allows wavelength changes without changing the direction of the reflected beam. The crystals are fabricated out of perfect Si or Ge single crystals which give according to the dynamical scattering theory [Batterman and Cole, 1964; Zachariasen, 1967] a wavelength smearing of

$$\Delta\lambda/\lambda = \sigma'_{iV} \cot\theta + \Delta G_{hkl}/G_{hkl} \quad , \tag{2.58}$$

which is, thus, determined by the beam divergence σ'_{iV} and the extension of the reciprocal lattice point (as caused by extinction effects). For $\sigma'_{iV} = 0.2$ mrad and the Ge(111) reflection this gives $\Delta\lambda/\lambda = 1.15 \times 10^{-3}$. The first monochromator crystal receives the total thermal load of the wiggler radiation that is typically around 200 Watt/mm^2 in multi-GeV storage rings [Spencer and Winick, 1980]. It, thus, has to be water-cooled in a sophisticated manner in order to avoid that lattice strains destroy the energy resolution (ΔG_{hkl}-term in (2.58)). In the future the "upstream"-crystal will presumably be cooled with liquid nitrogen [Bilderback, 1986; Freund and Marot, 1991]. The choice of the (111) reflection has the practical advantage that the usually strong first harmonic ($\lambda/2$) is almost completely suppressed, since the (222) reflection in a diamond lattice is quasi-forbidden [Roberto et al., 1974], $|F_{222}|^2/|F_{111}|^2 \cong 10^{-5}$. The higher harmonics $(\lambda/3, \ldots)$ which can pass through the monochromator system are eliminated subsequently by means of total external reflection from a Au coated mirror which is also in many cases (as at this beamline) the optical element which focuses the beam to a small spot at the sample position. The toroidal mirror at the W1 station is doubly focusing and has therefore two radii of curvature, at this station $r_T = 3.2$ km for the vertical (*"tangential"*) and $r_S = 15.5$ cm for the horizontal (*"sagittal"*) focusing. The two associated focal lengths f_S and f_T of such a device are [Sparks et al., 1980]

$$2f_S = \frac{r_S}{\cos \alpha_M} \quad \text{and} \quad 2f_T = r_T \cos \alpha_M \qquad (2.59)$$

and thus only equal for $\cos^2 \alpha_M = r_S/r_T$, or (in our case) for $\alpha_M = 6.96$ mrad. For all other settings of the mirror the beam focus is accordingly smeared out. The toroidal mirror is installed close to the halfway position between the source and the sample and produces then a 1 : 1-image of the source point at the sample position (see the "distorted" ruler at the bottom of Fig.2.14). The decision whether the monochromator system is placed up- or downstream of the mirror depends on the energy resolution requirements of the experiments; since the focusing mirror increases the divergence of the beam, the optimum spectral purity of the beam is achieved with the monochromator in front of the mirror as in Fig. 2.14. It should be clear that the above discussion of the performance of the x-ray optical elements touches only very elementary issues. Today sophisticated methods in conceiving new optical devices are applied which range from new analytical phase space methods [Matsushita and Kaminaga, 1980] to numerical ray tracing methods. The proper discussion of these concepts which have emerged with the outstanding properties of synchrotron radiation would warrant a review of its own.

A considerable experimental complication of surface investigations is the special requirement for the sample environment: The surface has to be kept in ultrahigh vacuum (UHV) throughout the experiment, if surface effects are to be studied on a clean surface. In the realisation of such UHV instruments two different experimental approaches have been pursued in the past:

a) In situ experimental stations, where the UHV preparation of the sample surface and the subsequent x-ray studies can be done within one huge UHV system [Brennan and Eisenberger, 1984; Fuoss and Robinson, 1984; Vlieg et al., 1987]. b) Portable UHV chambers which can be plugged onto a big sample preparation unit and then transferred with the sample onto the surface diffractometer [Robinson, 1983; Johnson et al., 1985]. Figure 2.15 depicts a schematic layout of a "portable baby chamber" which has been in use for many years at HASY-LAB [Johnson et al., 1985]. A "2π"-access of the x-ray beam onto the sample surface is enabled by a dome-shaped Be window. During a two weeks scattering experiment rest gas (predominantly H_2) will be adsorbed at the sample surface.

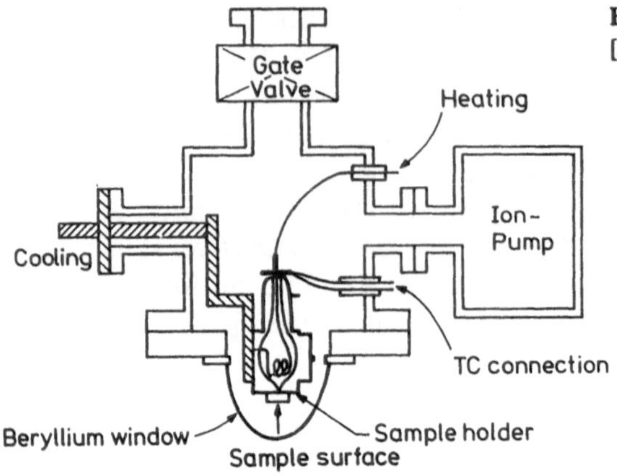

Fig. 2.15. Portable UHV x-ray chamber [Johnson et al., 1985]

The *"monolayer time"* τ_{ML} (time required for a monomolecular layer to form on a gas-free surface) can be estimated from the molecular arrival rate to be [King, 1979; Vaughan, 1986] $\tau_{ML} = 3.2 \times 10^{-6}/p$ (p in mbar), which gives $\tau_{ML} \simeq 12$ hours for gas pressures of $p = 10^{-10} - 10^{-11}$ mbar and for a sticking coefficient assumed to be 1.

2.4.4 Sample Preparation and Characterization

Prior to the actual surface scattering experiment the preparation of an atomically clean single crystal surface is required consisting of the mechanical polishing and the subsequent cleaning under UHV conditions. The general problem is that — though there are some generally valid rules for any sample surface — the experimentalist is forced either to develop his own sample-specific technique for a given surface or to hunt for recipies in the literature. The latter often means tracing the origin of a preparation procedure through a pile of publications. For glancing angle (x-ray or neutron) scattering experiments the proper mechanical polishing of the surface is usually one of the most time-consuming steps: It has to be assured that the surface

a) is macroscopically flat over a large area of typically 5×5 mm^2,
b) has a small surface roughness on the scale of the x-ray wavelength,
c) has homogeneous lateral crystal properties and
d) is exactly aligned with respect to the lattice planes (typically $|\Delta\phi| < 3$ mrad).

The surfaces of "soft" crystals (as Pb, Al or Cu) generally develop the socalled *"cushion effect"* during polishing which means the severe rounding at the edges of the surface. This can usually be minimized by embedding the surface into a mask (preferentially of the same material). During the polish (typically with grain sizes between 20 μm and 0.1 μm) the sample weight should be completely suspended. A final ($\simeq 1$ min) treatment of the surface with a 50 nm SiO$_2$ suspension produces excellent surfaces in the case of Fe$_3$Al(110), Cu$_3$Au(100), Al(110) and other metal surfaces (inkl. W, Ni and Cu). However, since this SiO$_2$ suspension is, as the *Syton* polish, chemically reactive, it should be used with care in the case of alloys, where heterogenous etching may occur. The same holds in particular for an electrochemical final treatment.

The subsequent UHV treatment consists of repeated Ar sputtering and annealing of the surface. With x-ray photospectroscopy (XPS), Auger analysis and LEED the progress of the cleaning procedure is controlled between the cycles. Successful UHV treatments of the various surfaces of elements have been collected by Musket et al. [1982].

As an example of the online glancing angle x-ray control of the mechanical polishing of a surface we refer to the experiments on $Fe_3Al(110)$ by Lied and coworkers [1989] described in Sect. 2.3. Here I add some remarks on Al(110) surfaces: It is indispensable that the sample surface is embedded in an Al mask and then *slowly and weightlessly* polished down to a grainsize of 50 nm [Dosch et al., 1991b]. After a conventional UHV treatment, however employing only a 500eV Ar sputtering (see [Jona et al., 1972]), which generated an atomically clean $(1 \times 1$ non-reconstructed) single crystal surface (as checked by XPS and LEED), the surface quality has been checked by evanescent Bragg scattering of x-rays. Figure 2.16 shows the specular beam profiles observed at two different grazing angles and the α_f-profile of the (002) Bragg reflection recorded for $\alpha_i/\alpha_c = 0.81$. The almost unbroadened specular beam indicates that the Al surface is mirrorlike. From the additional analysis of the Bragg profile within the DWBA (full line in Fig. 2.16b) one follows that the surface waviness is $\Delta n_s = 0.50$ mrad, the surface roughness is $\varrho = 10 \pm 2$ Å and the misalignment of the Bragg planes with respect to the surface is $\Delta \phi = -2.4$ mrad.

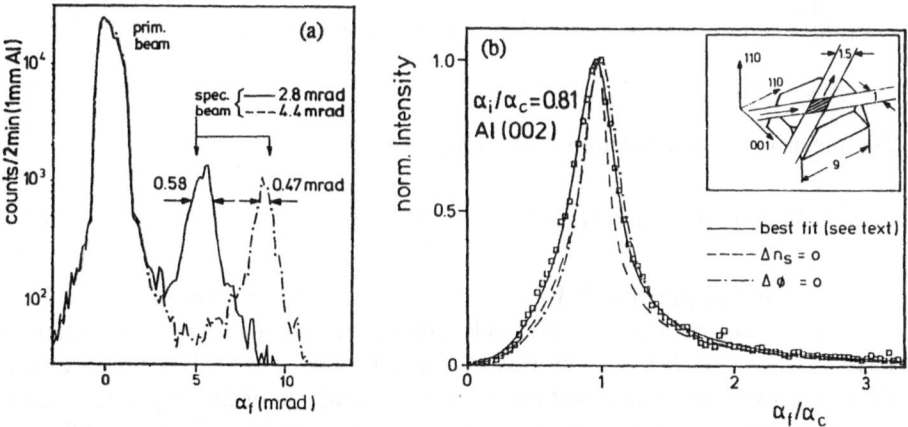

Fig. 2.16. α_f-profiles (a) of the specular beam (for two incidence angles α_i) and (b) of the (002) Bragg intensity observed at the Al(110) surface after the final UHV treatment [Dosch et al., 1991b]

3. Evanescent Neutron Scattering

3.1 Neutron Index of Refraction

In the previous section we derived the evanescent x-ray scattering phenomena from Maxwell's equations. In analogy we start up the neutron case with the (stationary) Schrödinger equation

$$\left(-\frac{\hbar^2}{2m}\nabla^2 + V(\boldsymbol{r})\right)\Psi_\sigma(\boldsymbol{r}) = E\Psi_\sigma(\boldsymbol{r}) \tag{3.1}$$

for the wave function $\Psi_\sigma(\boldsymbol{r})$ of a neutron with mass m and spin σ which experiences a spatially varying potential $V(\boldsymbol{r})$ (for a review see [Koester and Steyerl, 1977; Sears, 1986]). $E = |\boldsymbol{p}|^2/2m = \hbar^2 k^2/2m$ is the kinetic energy of the neutron while it travels in vacuum. The formal analogy between the Schrödinger equation and the Helmholtz equation (1.1) can be established by rearranging (3.1) in the form

$$\left(\nabla^2 + n^2(\boldsymbol{r})k^2\right)\Psi_\sigma(\boldsymbol{r}) = 0 \tag{3.2}$$

with the neutron index of refraction $n(\boldsymbol{r})$ which follows from (3.1,2) to be

$$n^2(\boldsymbol{r}) = 1 - V(\boldsymbol{r})/E \quad . \tag{3.3}$$

The mean index of refraction n is then given by

$$n = N \int d^3\boldsymbol{r} \left(1 - V(\boldsymbol{r})/E\right)^{1/2} \simeq 1 - V_0/2E \tag{3.4}$$

with the *"optical potential"* $V_0 \equiv N \int V(\boldsymbol{r})\, d^3\boldsymbol{r}$ and the number density N of the atoms. The for our purposes relevant neutron-matter interaction potential $V(\boldsymbol{r}) = V_n(\boldsymbol{r}) + V_m(\boldsymbol{r})$ is composed out of two parts, the nuclear interaction of the neutron with the nuclei and the magnetic interaction of the magnetic moment of the neutron with the magnetic moment of the unpaired electrons. All other interactions, due to gravity [Maier–Leibnitz, 1962][1], charge-interactions [Foldy, 1955] and the Schwinger-interaction [Schwinger, 1948] can savely be neglected. I will shortly discuss the features of $V_n(\boldsymbol{r})$ and $V_m(\boldsymbol{r})$ in the following:

---

[1] The gravitational drop of neutrons in a monochromatic beam was first demonstrated by McReynolds [1951a,b] at Brookhaven and later pioneered by Maier–Leibnitz in Munich.

a) $V_n(r)$: All one knows today is that this interaction is attractive (strong interaction) and short ranged (about 10^{-15} m). From neutron scattering experiments, however, one has deduced the scattering lengths $b_\pm$ which govern the scattering of the neutron from one single nucleus. The index $\pm$ indicates that the strength of $V_n(r)$ and consequently the value of $b_\pm$ depend on the neutron spin orientation ($\pm$) with respect to the nuclear spin. In order not to complicate the subsequent discussion we don't consider the consequences of spin-dependent nuclear scattering and refer the reader instead to standard textbooks as by Bacon [1975], Squires [1978], Marshall and Lovesey [1971] and others. So we drop the index $\pm$ and discuss the coherent scattering length b_c. With the above information one can construct a pseudopotential

$$V_n(r) = \frac{2\pi\hbar^2}{m_n} \sum_i b_{ci}\delta(r - R_i) \tag{3.5}$$

which reproduces the observed value b_{ci} associated with the nucleus at R_i within the first Born approximation (*"Fermi pseudopotential"*). Apart from a few exceptions (like H, Mn and Ti) b_c is positive, thus we have now obtained a repulsive potential which we use to describe the actual attractive interaction. This fact may cause confusion, when one forgets about the "pseudo" in the artificial potential (3.5). From (3.5) we deduce the nuclear optical potential

$$V_{n0} = \frac{2\pi\hbar^2}{m_n} N_n \langle b_c \rangle \tag{3.6}$$

with the average coherent scattering length $\langle b_c \rangle = \sum N_i b_{ci}$ (*"nuclear jellium"*) and the nuclear number density N_n.

b) $V_m(r)$: With its spin σ_n the neutron carries an anomalous (since it is uncharged) magnetic moment $\mu_n = \gamma\mu_n\sigma_n$, where $\mu_n = e\hbar/2m_p$ (= 3.152×10^{-8} eV/Tesla) is the nuclear magneton[2] and $\gamma = -1.9130$ the neutron gyromagnetic ratio. When electrons in partially filled shells provide a spatially varying magnetic field $B(r)$, the incoming neutron experiences a magnetic interaction

$$V_m(r) = -\mu_n \cdot B(r) \tag{3.7}$$

with $B(r) = \text{rot}\, A(r)$ and the vector potential $A(r) = (M \times r)/r^3$, thus, after a simple algebra,

$$B(r) = \mu_0 M(r) + \frac{\mu_0}{4\pi} \int_{V_+} d^3r' \, (M(r') \cdot \nabla')\nabla' \frac{1}{|r - r'|} \quad , \tag{3.8}$$

where $M(r)$ denotes the local magnetization density and V_+ the half space filled with matter (see Fig. 3.1),

$$M(r) = \sum_e \mu_e \delta(r - R_e) \tag{3.9}$$

---

[2] In most textbooks the magneton is given in cgs units: $\mu = e\hbar/2mc$.

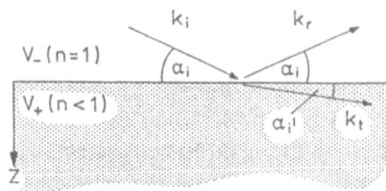

Fig. 3.1. Neutrons impinging onto an interface at $z = 0$: V_- is the vacuum half space, V_+ the half space filled with matter; k_i, k_r, k_t are the incident, reflected and transmitted wave vectors

with $\mu_e = -2\mu_B s_e$ being the magnetic moment associated with the electronic spin s_e $(|s_e| = \pm 1/2)$ at R_e and $\mu_B = e\hbar/2m_e$ (= 5.788×10^{-5} eV/Tesla) the Bohr magneton. For the sake of simplicity we have neglected the contributions from the orbital motions[3] in (3.9).

While today the form of $V_m(r)$ is common wisdom, it was the subject of a considerable controversy in the 50ies [Eckstein, 1949, 1950; Lax, 1950, 1951; Halpern, 1949, 1952; Hughes and Burgy, 1951] based upon open questions about the origin of the neutron magnetic moment [Bloch, 1936; Schwinger, 1937]. I strongly encourage the interested reader to have a closer look into these arguments. Here we proceed by deriving the magnetic optical potential V_{m0} which follows from (3.7-9) after remembering that $\int B(r)\, d^3r = \mu_0 \langle M \rangle$ and

$$\langle M \rangle = -2N_m \mu_B \langle s_e \rangle \tag{3.10}$$

to be

$$V_{m0} = -\langle \mu_n \cdot B \rangle = 2N_m \gamma \mu_0 \mu_n \mu_B \langle \sigma_n \cdot s_e \rangle = \frac{2\pi\hbar^2}{m_p} N_m \gamma r_e \langle \sigma_n \cdot s_e \rangle \tag{3.11}$$

where N_m is the number density of the magnetic atoms and $\langle s_e \rangle$, $\langle M \rangle$ the mean electronic spin and the mean magnetization of the sample, respectively, as averaged within the coherence length of the neutron beam.

Now we have all ingredients to assemble the average neutron index of refraction for a transparent medium (no absorption): Inserting (3.6) and (3.11) into (3.4) we get

$$n_\pm \equiv 1 - \delta_\pm = 1 - \frac{\lambda^2}{2\pi} \left(N_n \langle b_c \rangle + N_m \gamma r_e \langle \sigma_n \cdot s_e \rangle \right) \tag{3.12}$$

with r_e as the classical electron radius.

Commonly in literature (see [Hughes, 1954; Felcher, 1981]) the magnetic part of $\delta_\pm$ is written in a different form, as $\pm N_m \mu C_1$ with $C_1 = -2.659 \times 10^{-5}$ Å and μ the number of Bohr magnetons per magnetic atom. By comparison with (3.12) the numerical constant C_1 is readily identified as $C_1 = (\gamma/2)r_e$. By denoting μC_1 the average magnetic scattering length $\langle b_m \rangle$, the index of refraction of polarized neutrons takes on the particularly marked form

---

[3] This assumption holds for 3d-transition metals and some magnetic insulators, where the orbital momenta are quenched (see [Stoner, 1929]).

$$n_\pm = 1 - \frac{\lambda^2}{2\pi} \left(N_n \langle b_c \rangle \mp N_m \langle b_m \rangle \right) \quad . \tag{3.13}$$

After this seemingly simple derivation of $n_\pm$ I want to stress at this point that the fact $\langle n \rangle \neq 1$ is a consequence of a multiple scattering phenomenon. Its rigorous calculation is a rather cumbersome task which has been attacked by many people in the past years [Foldy, 1945; Lloyd and Berry, 1967; Sears, 1982; Warner and Gubernatis, 1985]. In the framework of a multiple scattering approach n reads (neglecting correlations between the nuclei)

$$n = 1 + \frac{2\pi}{k^2} N \lim_{\theta \to 0} f(\theta) \tag{3.14}$$

with $f(\theta) = -b_c + i k b_c^2 + O(k^2)$ (see e.g. [Bethe and Morrison, 1956]) as the scattering amplitude observed in the scattering angle θ. Since $b_c = O(r_n) \simeq 5\,\mathrm{fm}$ (r_n is the linear dimension of the nucleus) we find for thermal neutrons $|k b_c| \simeq 10^{-4}$ and recover expression (3.13) for the case $\langle b_m \rangle = O$.

So far we did not account for the attenuation of the neutron beam in matter. Assume now a total interaction cross section σ_t per atom, then the neutron intensity decays as

$$I(z) = I_0 e^{-N_n \sigma_t z} \quad . \tag{3.15}$$

Such absorption experiments have been performed for many years in order to determine σ_t (for a review see [Adair, 1950]). From a phenomenological point of view we would use a complex index of refraction $n \equiv 1 - \delta + i\beta$ with the extinction coefficient $\beta = \mu\lambda/4\pi$,

$$I(z) = |A_0 e^{inkz}|^2 = I_0 e^{-2\beta kz} \quad . \tag{3.16}$$

By comparing (3.15) with (3.16) we arrive at

$$\beta = \frac{N_n}{2k} \sigma_t = \frac{N_n}{2k} \left(\sigma_{inc} + \sigma_a \right) \quad , \tag{3.17}$$

where the cross sections σ_{inc} and σ_a account for the incoherent scattering and nuclear reactions, respectively[4]. On the other hand, since $\beta = \mathrm{Im}\{n\}$, we get together with (3.14) a famous general relation,

$$\beta = \frac{2\pi}{k^2} N_n \mathrm{Im}\{f(0)\} = \frac{N_n}{2k} \sigma_t \quad ,$$

or,

$$\sigma_t = \frac{4\pi}{k} \mathrm{Im}\{f(0)\} \tag{3.18}$$

which is customarily known as the *"Bohr–Peierls–Placzek relation"* or the *"optical theorem"* and relates the total dissipative cross section to the imaginary part

---

[4] In the presence of strong Bragg scattering σ_{coh} also adds to the attenuation of the transmitted beam ("Renninger effect").

of the forward scattering amplitude (for a rigorous quantum-theoretical discussion of (3.18) see e.g. Sakurai [1985] and also Feenberg [1932]). The complete complex index of refraction for a polarized neutron beam reads with (3.13) and (3.17)

$$n_\pm \equiv 1 - \delta_\pm + i\beta = 1 - \frac{\lambda^2}{2\pi}\left(N_n\langle b_c\rangle \mp N_m\langle b_m\rangle\right) + i\frac{\lambda}{4\pi}N_n\left(\sigma_{inc} + \sigma_a\right). \quad (3.19)$$

We will see in the following that neutron scattering studies performed under the condition of total external reflection suffer from the very low intensity. Unfortunately no new neutron source comparable to the synchrotron radiation is in sight. As a consequence one would use neutrons instead of x-rays only in those cases where the trade-off qualities of evanescent neutron waves are required, these are the magnetic interaction which allows the study of surface magnetism[5] and the very low absorption cross section which gives access to buried interfaces. We will focus in the following on these two points, but will derive first the equations which govern neutron optics.

3.2 Elementary Neutron Optics

Let us assume that we have a medium with $\delta_\pm$ greater than 0. Then total external reflection of neutrons occurs when a monochromatic neutron beam with wavelength λ impinges onto a mirror-like surface of this substance at an angle α_i which is less than the associated critical angle (Fig. 3.1)

$$\alpha_{\pm c} = \left(2\delta_\pm\right)^{1/2} = \lambda\left[\left(N_n\langle b_c\rangle \mp N_m\langle b_m\rangle\right)/\pi\right]^{1/2} . \quad (3.20)$$

The critical angle $\alpha_{\pm c}$ of a substance can be determined quite accurately. This in turn allows a precise experimental determination of the coherent scattering length $\langle b_c\rangle$ of solids and liquids [Koester, 1965, 1967] and even of gases[6].

We proceed by deriving the neutron reflection and transmission coefficients of the solid-vacuum interface via the ansatz

$$\Psi(r) = \begin{cases} e^{ik_{i\parallel}\cdot r_\parallel}\left(e^{ik_{iz}z} + R_i e^{-ik_{iz}z}\right) & \text{in } V_- \\ e^{ik_{i\parallel}\cdot r_\parallel} T_i e^{ik'_{iz}z} & \text{in } V_+ \end{cases} , \quad (3.21)$$

for the neutron wave function $\Psi(r)$ (we dropped the index σ), where k_{iz}, k'_{iz} and $-k_{iz}$ are the z-components of the wave vectors of the incident, transmitted and reflected neutron waves which have in the coordinate system of Fig. 3.1 the same form as in the x-ray case discussed before $\left(\text{see } (2.7,10)\right)$. In particular

---

[5] Despite of the low magnetic photon scattering cross section magnetic x-ray scattering from surfaces appears feasible with the advent of new low-emittance synchrotron sources.

[6] The fact that the presence of a gas phase changes the critical angle of the solid interface can be used to gauge the coherent scattering length of a gas by measuring the gas-pressure dependence of the reflected neutron intensity [McReynold, 1951a,b].

$$k'_{iz} = k''^{(\pm)}_{iz} = k\left(\sin^2\alpha_i - 2\delta_{\pm} + 2i\beta\right)^{1/2} \equiv k\,\sin r_{\pm}\,\alpha_i \quad . \tag{3.22}$$

We determine the coefficients R_i and T_i from the boundary conditions

$$\Psi_i(r) + \Psi_r(r)\big|_{z=0} = \Psi_t(r)\big|_{z=0} \quad ,$$

$$\frac{\partial\left(\Psi_i(r) + \Psi_r(r)\right)}{\partial z}\bigg|_{z=0} = \frac{\partial\Psi_t(r)}{\partial z}\bigg|_{z=0} \tag{3.23}$$

and obtain $1 + R_i = T_i$ and finally

$$T_i = \frac{2k_{iz}}{k_{iz} + k'_{iz}} \quad , \qquad R_i = \frac{k_{iz} - k'_{iz}}{k_{iz} + k'_{iz}} \quad . \tag{3.24}$$

Figure 3.2 shows calculated neutron reflectivity curves of some standard materials: Si and Gd as examples of systems with negligible and strong absorption, respectively, Ni^{58} which is used as coating in neutron guides[7]. For details concerning neutron optics I strongly recommend the excellent textbook by Sears [1989].

One of the trade-off qualities of neutrons, the energy and wavelength selectivity via a time-of-flight (TOF) analysis, allows the conception of particularly simple and efficient reflectometers which work at a fixed incident angle and tune the wavelength of the neutron radiation by scanning the neutron velocity along a course set by an incident beam monitor and the final detector. Since this method necessitates a chopped neutron beam, it is no accident that two of the so far most efficient reflectometers are installed at pulsed neutron sources (spallation

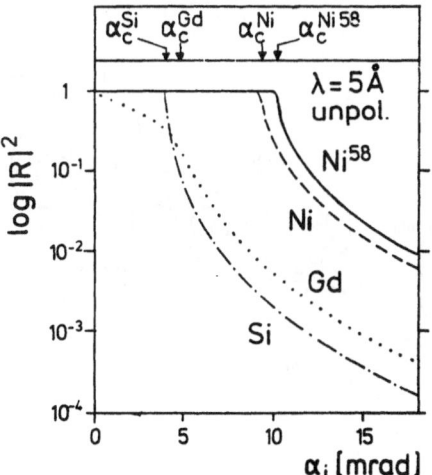

Fig. 3.2. Neutron reflectivity curves for various materials as a function of α_i using monochromatic neutrons with wavelength $\lambda = 5\,\text{Å}$. (Note the logarithmic scale)

---

[7] Ni^{58} is an $S = 0$ state, therefore $\sigma_{inc} = 0$.

sources), whereas at continuous neutron sources reflectometry has been treated somewhat stepmotherly in the past.

Several dedicated neutron reflectometers are operating in various neutron scattering facilities, as e.g.

— the polarized neutron reflectometer (POSY) at spallation source IPNS of the Argonne National Laboratory in Argonne, Illinois [Felcher et al., 1987]: fixed α_i with TOF analysis

— the critical reflection spectrometer (CRISP) at the spallation source ISIS of the Rutherford Appleton Laboratory in Didcot, UK [Penfold et al., 1987]: fixed α_i with TOF analysis

— the total reflection machine (TOREMA) at the research reactor of the Forschungszentrum Jülich [Stamm et al., 1989]: fixed λ with α_i-scan

— the evanescent wave spectrometer (EVA) at the high flux research reactor of the Institut Laue Langevin in Grenoble [Al Usta et al., 1991]: fixed λ with α_i-scan; dedicated to evanescent neutron Bragg scattering.

As a typical TOF reflectometer we shortly discuss the schematic diagram of the CRISP instrument (Fig. 3.3) at the pulsed source ISIS near London: The instrument views the 20 K hydrogen moderator which provides a white beam with wavelengths associated with the 20 K Maxwell spectrum. The actually used wavelength spread is defined by a single disc chopper and by socalled "frame overlap" mirrors which reflect very slow neutrons ($\lambda \geq 13$ Å) out of the beam. A standard (fixed) incidence angle is $\alpha_i = 1.5°$, thus, the reflectometer ranges a perpendicular momentum transfer from $Q_z = 4 \times 10^{-3}$ Å^{-1} to 0.65 Å^{-1} by a simple TOF analysis of the detected neutrons without any moving parts. The collimation of the beam is realized by coarse collimating jaws (which shield unuseful neutrons) followed by two vertical slits (S1,2) which taylor the vertical

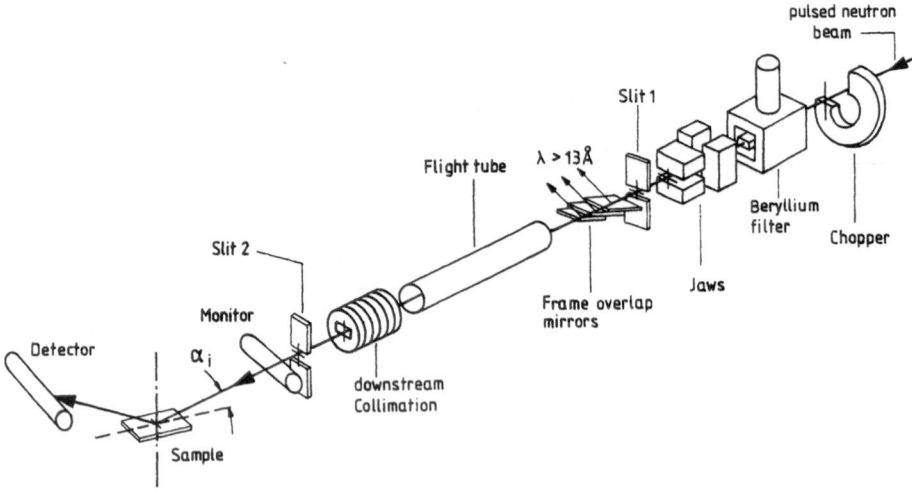

Fig. 3.3. Neutron reflectometer CRISP at the spallation source ISIS [Penfold et al., 1987]; for explanation see text

beam size (0.25 mm − 6 mm) and divergence to the sample dimensions. The detector (He³ single detector or a linear position sensitive detector) is placed approximately 1.5 m away from the reflecting sample surface. When surface ferromagnetism is studied, the incident neutrons are polarized.

3.2.1 Magnets and Superconductors

According to (3.22) the z-component of the perpendicular wavevector component in a magnetic medium has two different values depending on the spin of the neutron with respect to the magnetization in the substance. This effect can be exploited in an elegant way in order to measure surface magnetization profiles via the total external reflectivities $R_{i\pm}$ [Felcher, 1981]. In order to demonstrate the effect, we show the observed reflectivity profiles $R_{i\pm}$ and the resulting "flipping ratio" $F_R \equiv R_{i+}/R_{i-}$ at a Ni(100) surface in the paramagnetic phase (350°C) (Fig. 3.4a) and in the ferromagnetic phase (250°C) (Fig. 3.4b). These data have recently been obtained at the POSY reflectometer of the Argonne National Lab-

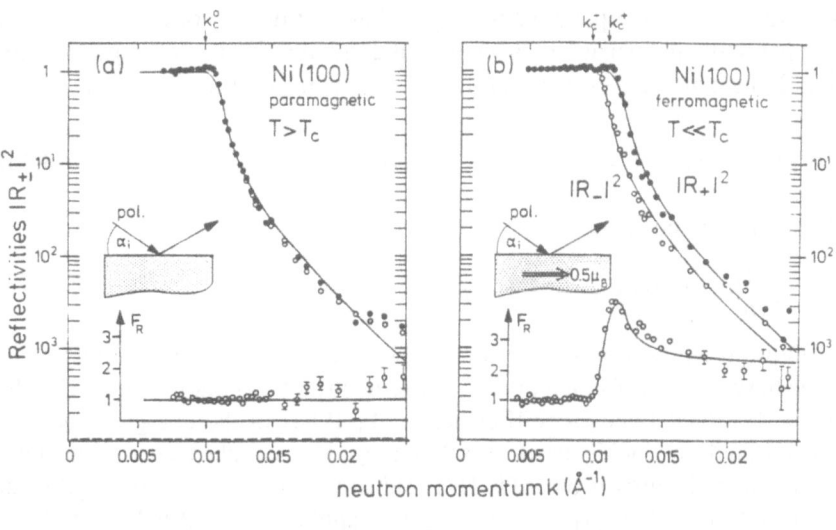

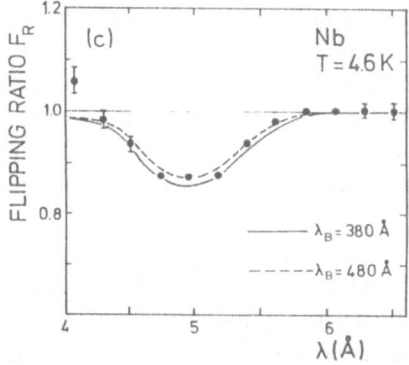

Fig. 3.4a-c. Reflectivity flipping ratios F_R of polarized neutrons: Measured reflectivities and F_R from the (100) Ni surface (a) in the paramagnetic and (b) in the ferromagnetic state [Dosch et al., 1991d]; (c) observed F_R versus λ at the surface of superconducting Nb exposed to an external magnetic field [Felcher et al., 1984]

oratory [Dosch et al., 1991d]. The calculated flippig ratio (bottom full line in Fig. 3.4b) assumes a ferromagnetic single domain with $\mu = 0.5\,\mu_B$ and agrees well with the observation.

An intriguing application of this scheme is the measurement of the penetration depth λ_b of an external magnetic field into a semi-infinite material as it undergoes a phase transition at T_{SC} into the superconducting state (Fig. 1.1). As a consequence of the Cooper pairing among the conduction electrons below T_{SC} the solid becomes an ideal diamagnet whose bulk ("$z = \infty$") is characterized by $B = 0$ (Meissner Ochsenfeld effect). At the surface ("$z = 0$") of the semi-infinite superconducting solid an applied external B-field decays exponentially fast into the medium, $B = B_0 \exp(-z/\lambda_b)$, where in socalled "clean" systems λ_b reads (see e.g. [Ashcroft and Mermin, 1976]),

$$\lambda_b = \left(\frac{m_e}{\mu_0 N_e^* e^2} \right)^{1/2} t^{-\nu} \quad . \tag{3.25}$$

The prefactor ("*London depth*") is typically between $400\,\text{Å}$ and $650\,\text{Å}$ for type-I superconductors, while within mean field theory $\nu_{\text{MFA}} = 1/2$ (see Chap. 4). In (3.25) t is the reduced temperature and N_e^* the number density of electrons which form Cooper pairs. Associated with a constant B-field in V_- and an exponentially decaying B-field in V_+ is a depth profile of the neutron refractive index

$$n_\pm(z) \equiv 1 - \delta_\pm = 1 - \frac{\lambda^2}{2\pi} \left(N_n \langle b_c \rangle \mp C_2 B_0 \left(1 - e^{-z/\lambda_b} \right) \right) \tag{3.26}$$

with the numerical constant

$$C_2 \equiv \frac{m_n \nu_n}{2\pi \hbar^2} = 1.2089 \times 10^{-6}\,\text{Å}^{-2}/\text{Tesla} \tag{3.27}$$

which follows directly from the definitions (3.4,7). The neutron reflectivity measurements have been performed at the intense neutron (spallation) source (IPNS) at Argonne National Laboratory [Felcher et al., 1984] in the energy dispersive mode (fixed angle of incidence $\alpha_i = 5.93$ mrad). The sample was a $5\,\mu\text{m}$ thick Nb film (epitaxially grown on a (111) Si substrate) which was investigated at some temperatures below $T_{SC} = 9.25$ K with an applied external field B-field of 500 Oe. A typical example of the observed flipping ratios is shown in Fig. 3.4c for $T = 4.6$ K as a function of the neutron wavelength. The full and the dashed line are the results of a numerical calculation with $\lambda_b = 380\,\text{Å}$ and $480\,\text{Å}$, respectively, i.e. this result lies in good agreement with the theoretical prediction $\lambda_b = 440\,\text{Å}$. Similar experiments have been reported on the high-T_c superconductor YBaCuO by Felici et al. [1987] and Mansour et al. [1989] with rather controversial results presumably due to differences in the sample preparation technique.

As one knows, also x-rays exhibit a tiny, but nonzero magnetic interaction cross section with matter (see e.g. [Platzman and Tzoar, 1970; de Bergevin and Brunel, 1981; Blume, 1985; Lovesey, 1987]), and it may puzzle the reader why this does not show up in the x-ray index of refraction. The answer to this lies in the nature of the x-ray magnetic interaction: The (for the experts: non-resonant) magnetic x-ray scattering amplitude of one unpaired electron at the origin with orbit momentum p_e and spin σ_e is given by

$$f_m(\theta) \sim -i \frac{Q \times p_e}{\hbar k^2} \cdot G_1 + \frac{1}{k^2} \sigma_e \cdot (G_1 + G_2) \quad , \tag{3.28a}$$

where

$$G_1 = e_f \times e_i \quad , \tag{3.28b}$$

$$G_2 = (k_f \times e_f) \cdot (k_f \cdot e_i) - (k_i \times e_i) \cdot (k_i \cdot e_f) \\ - (k_f \times e_f) \times (k_i \times e_i) \tag{3.28c}$$

and $Q = k_f - k_i$. In forward scattering ($\theta = 0$, i.e. $k_f = k_i$ and $e_f = e_i$) this expression vanishes exactly, $f_m(0) = 0$, which entails, because of (3.14), a zero magnetic contribution to the x-ray index of refraction. At absorption edges, where resonant magnetic scattering occurs (see [Hannon et al., 1988; Isaacs et al., 1989]), a small magnetic contribution to the x-ray refractive index emerges. Future experiments will tell, whether this subtle effect can be used to extract interfacial magnet moments.

3.2.2 Buried Interfaces

We consider now the situation of the surface of a semi-infinite solid (2) which is covered by a matter (1) of thickness Y as illustrated in Fig. 3.5a.

A neutron wave Ψ_0 with wave vector z-component k_{0z} impinges from the vacuum side onto this system and undergoes reflection, refraction and scattering

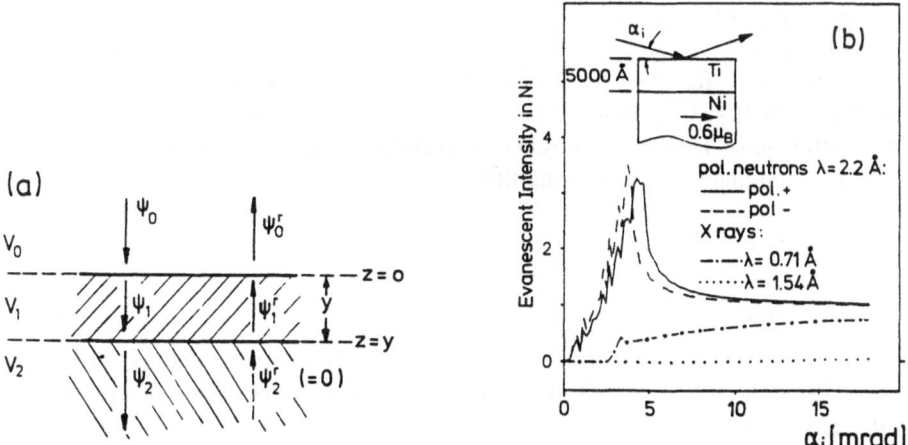

Fig. 3.5a,b. Evanescent neutrons at a buried interface in the case of ferromagnetic Ni coated with Ti. (a) Wave fields in the halfspaces V_0 (*vacuum*) and V_2(Ni) and in the coating V_1(Ti); (b) transmitted intensities at the buried Ni interface for polarized neutrons compared with x-rays (Cu- and Mo-radiation)

effects. We will discuss the form of the reflectivity and scattering as observed in V_-. The continuity requirements for the wave functions at the two interfaces at $z = 0$ and $z = Y$ are (note that $\Psi_2^r = 0$ is the reflected wave from the interface at ∞)

$$z = 0: \quad \Psi_0 + \Psi_0^r = \Psi_1 + \Psi_1^r \quad , \qquad k_{0z}(\Psi_0 - \Psi_0^r) = k_{1z}'(\Psi_1 - \Psi_1^r) \tag{3.29}$$

$$z = Y: \quad \Omega_1 \Psi_1 + \Omega_1^r \Psi_1^r = \Psi_2 \quad , \qquad k_{1z}'(\Omega_1 \Psi_1 - \Omega_1^r \Psi_1^r) = k_{2z}' \Psi_2 \quad , \tag{3.30}$$

with the phase factors $\Omega_1 = \exp(ik_{1z}' Y)$, $\Omega_1^r = \exp(-ik_{1z}' Y)$ occuring at $z = Y$ (as usual, primed wave vectors are complex). The solution is simply found to be

$$\Psi_1^r = \Psi_2 \frac{k_{1z}' - k_{2z}'}{2k_{1z}' \Omega_1^r} \quad , \qquad \Psi_1 = \Psi_2 \frac{k_{1z}' + k_{2z}'}{2k_{1z}' \Omega_1} \quad , \qquad \Psi_0^r = \Psi_1 + \Psi_1^r - \Psi_0$$

$$\Psi_2 = \Psi_0 \frac{4k_{0z} k_{1z}'}{\Omega_1^r(k_{0z} + k_{1z}')(k_{1z}' + k_{2z}') + \Omega_1(k_{0z} - k_{1z}')(k_{1z}' - k_{2z}')} \tag{3.31}$$

and allows now the calculation of the reflected amplitude (Ψ_0^r) , the amplitude (Ψ_2) in the semi-infinite matter (V_2) and the amplitude ($\Psi_1 + \Psi_1^r$) in the finite matter (V_1). Rearrangement of (3.31) then gives the transmission and reflection coefficients, T_{02} and R_{02}, of the total assembly to be

$$T_{02} \equiv \Psi_2 / \Psi_0 = \frac{T_{01} T_{12} \, e^{ik_{1z}' Y}}{R^y} \quad , \tag{3.32}$$

$$R_{02} \equiv \Psi_0^r / \Psi_0 = \frac{R_{01} + R_{12} \, e^{2ik_{1z}' Y}}{R^y} \tag{3.33}$$

with

$$R^y = 1 + e^{2ik_{1z}' Y} R_{01} R_{12} \quad , \tag{3.34}$$

where T_{ab} and R_{ab} are the Fresnel coefficients (3.24) of the interface which separates V_b ($b = 1, 2$) from V_a ($a = 0, 1$). The derived Fresnel coefficients allow the inclusion of interfacial roughness: Following Vidal and Vincent [1984] one gets the roughness-modified quantities

$$T_{ab}^\varrho = T_{ab} e^{\frac{1}{2}(k_{bz} - k_{az})^2 \varrho^2} \tag{3.35}$$

$$R_{ab}^\varrho = R_{ab} e^{-2k_{bz} k_{az} \varrho^2} \quad , \tag{3.36}$$

where ϱ measures the roughness of the interface between V_b and V_a. For a nice early discussion of reflectivity from rough stratified media see Abeles [1950] and Heavens [1955].

Of course the derived expressions hold as well for x-rays (see [Born and Wolf, 1980]). The advantage of neutrons come into play, when the thickness Y of the coating becomes large (say ≥ 5000 Å): Then the comparatively strong imaginary part of the x-ray wave vector leads to $\exp(ik_{1z}'Y) = 0$ and thus to

$T_{02} = 0$ and $R_{02} = R_{01}$, i.e., the substrate is invisible for x-rays. When neutrons are used two effects unique for neutrons can be exploited:

a) When the coating has a negative coherent scattering length, no total reflection occurs at the first interface and no evanescent waves develop in the coating.
b) The absorption in the coating layer is generally small enough to allow the neutron wave to penetrate down to the buried interface which we are interested in.

With evanescent neutrons one could come into the stage to investigate in detail interfacial ferromagnetism and antiferromagnetism or, most fascinating, the spatial transition from ferromagnetic to antiferromagnetic ordering across an interface (see e.g. [Saurenbach et al., 1988]). As one typical example we envisage the surface of ferromagnetic Ni ($b_c = 1.03$ fm, $\mu = 0.6\mu_B$) coated with 5000 Å Ti ($b_c = -0.33$ fm) and calculate the evanescent intensity implanted underneath the Ni surface as a function of α_i in the case of polarized neutrons ($\lambda = 2.2$ Å) and x-rays ($\lambda = 0.71$ Å and 1.54 Å). Fig. 3.5b shows the normalized evanescent neutron intensity I_2/I_0 inside Ni which can be used in a grazing angle scattering experiment to extract structural and magnetic properties of the buried Ni surface (Sect. 3.3). While the (polarized) neutrons display the almost unperturbed transmission function within Ni which depends nicely on the polarization[8], the x-ray transmission into Ni is only poor for high-energetic Mo K_α radiation and virtually zero for Cu K_α radiation. (Normally the x-ray transmission is in addition insensitive to the magnetization of Ni). The potential of evanescent neutron scattering for the investigation of such intriguing problems which have considerable scientific and technical interest cannot be overestimated (for a recent review of magnetism at interfaces see [Shinjo, 1991]). As we will see in the next section, evanescent Bragg scattering is already feasible with neutrons. A related problem, a thin Co layer sandwiched between Cu layers has been studied by reflectivity measurements using polarized neutrons [Pescia et al., 1987].

3.3 Grazing Angle Neutron Scattering in DWBA

In order to derive the scattering cross section in the DWBA it is quite helpful to start with the conventional bulk cross section $d\sigma/d\Omega$ for elastic scattering: Fermi's Golden rule gives $d\sigma/d\Omega$ in the first Born approximation (see e.g. [Lovesey, 1984])

$$\frac{d\sigma}{d\Omega}(Q) = \left(\frac{m}{2\pi\hbar^2}\right)^2 \left|\langle \Psi_f | \Delta V | \Psi_i \rangle\right|^2 \tag{3.37}$$

with the amplitude

---

[8] The intensity wiggles for small α_i are thickness fringes from the Ti coating. Their observation, however, would require an unrealistically high grazing angle resolution.

$$\langle \Psi_f | \Delta V | \Psi_i \rangle = \int d^3 \boldsymbol{r} \, \Psi_f^* \Delta V \Psi_i \equiv V^{fi} \quad . \tag{3.38}$$

ΔV is the spatially varying part of the neutron-matter interaction potential which is composed of the nuclear and the magnetic part as discussed above, $\Psi_i(\boldsymbol{k}_i)$ and $\Psi_f(\boldsymbol{k}_f)$ are the wave functions of the incoming and scattered neutron, respectively, which are plane waves $\exp(i\boldsymbol{k}_{i,f} \cdot \boldsymbol{r})$. Thus, (3.38) simply becomes the Fourier transform of the neutron-matter interaction potential.

When the scattering is observed under the condition of total external reflection, the neutron waves are no longer plane waves in the halfspace V_+ (Fig. 3.1), but instead distorted waves as described by the transmission coefficient of the surface and by the complex wave vectors. We have to evaluate (3.38) in the two halfspaces $V_\pm$ with the associated Fresnel solutions (3.21) of the neutron wave functions.

It should be clear without noting that all the kinematic scattering effects from real surfaces which have been discussed in the context of grazing angle x-ray scattering hold as well for evanescent neutron scattering and will not be repeated in this chapter.

3.3.1 Nuclear Scattering and Surface Roughness[9]

The semi-infinite scattering cross section can be derived in a very close analogy to the x-ray case. The nuclear scattering potential $\Delta V_n = V_n - V_{n0}$ is only nonzero in V_+ and gives accordingly

$$V_n^{fi} = \langle \Psi_f | \Delta V_n | \Psi_i \rangle = \frac{2\pi \hbar^2}{m} \beta_+^{fi} \int_{V_+} \Delta \varrho_n(\boldsymbol{r}') e^{-i\boldsymbol{Q}' \cdot \boldsymbol{r}'} d^3 \boldsymbol{r}' \quad , \tag{3.39}$$

with $\beta_+^{fi} = T_f^* T_i$ and $\Delta \varrho_n(\boldsymbol{r}) = \sum_i (b_{ci} - \langle b_c \rangle) \, \delta(\boldsymbol{r} - \boldsymbol{R}_i)$. The square of the semi-infinite integral is the kinematic structure factor $S(\boldsymbol{Q}')$. From (3.39) we follow that the scattered intensity is rather simply

$$I(\boldsymbol{Q}') = I_0 (T_f^* T_i)^2 S(\boldsymbol{Q}') \tag{3.40}$$

which is fully analogous to $I_\parallel(\boldsymbol{Q}')$ (2.33) in the x-ray case (apart from the different underlying interaction).

We will discuss the implications of (3.40) in the case of diffuse neutron scattering from a rough surface. This issue is of considerable practical importance for the performance of neutron guide tubes, because surface roughness of the walls of the guide tube leads to a loss of the "guided" neutron intensity. Therefore it is no surprise that one of the first theoretical and numerical studies of surface roughness effects have been reported in this context [Steyerl, 1972], meanwhile, however, more modern approaches based on the DWBA have appeared [Sinha

---

[9] Here a surface is called rough in contrast to a mathematical surface and not in the thermodynamic sense as discussed in Sect. 5.3.

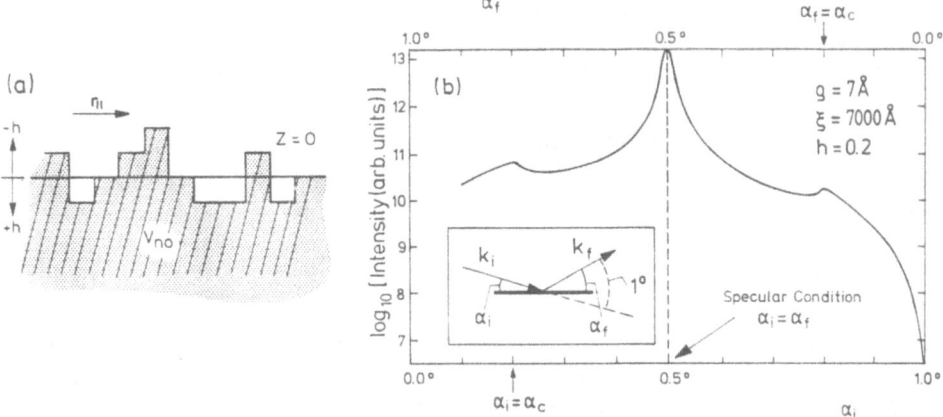

Fig. 3.6a,b. Evanescent neutron scattering from a rough surface [Sinha et al., 1988]: (a) derivation of scattering potential (see text); (b) diffuse scattering in forward direction along Q_z for varying α_i and a fixed scattering angle $2\theta = 1°$ (resulting in an α_f variation as shown in the top scale). Due to the large correlation length $\xi = 7000$ Å the diffuse intensity is distinctly peaked at the specular condition ($\alpha_i = \alpha_f$). For $\alpha_i = \alpha_c$ and $\alpha_f = \alpha_c$ the surface optics create the *Yoneda–Warren wings*

et al., 1988] which I will sketch briefly: At first we construct the perturbation potential $\Delta V_n(r_\parallel)$ which gives rise to the roughness-diffuse scattering: From Fig. 3.6a which shows the total nuclear potential of a rough surface we obtain $\Delta V_n(r_\parallel)$ after subtracting the nuclear jellium potential V_{n0} (3.6) associated with an average smooth surface at $z = 0$ in the form

$$\Delta V_n(r_\parallel) = -V_{n0}\,\mathrm{sgn}\big(h(r_\parallel)\big) \qquad \text{for } z \in [0, h(r_\parallel)] \tag{3.41}$$

with sgn(x) = ± 1 for $x \lessgtr 0$ and = 0 for $x = 0$. The evanescent diffuse scattering from (3.41) has to be calculated from (3.37), where Ψ_i is given by (3.21) and Ψ_f by the time-reversed solution

$$\Psi_f(r) = e^{ik_{f\parallel}\cdot r_\parallel} \begin{cases} e^{ik_{fz} z} + R_f^* e^{-ik_{fz} z} & \text{in } V_- \\ T_f^* e^{ik'_{fz}} & \text{in } V_+ \end{cases} . \tag{3.42}$$

The apparent complication in performing the bracket $\langle \Psi_f | \Delta V_n | \Psi_i \rangle$ is that Ψ_i and Ψ_f have a different form above and below the average smooth surface. This leads to cross terms similar to the ones which will be discussed below in the context of magnetic scattering (see Sect. 3.3.2). Sinha et al. [1988] proposed to approximate the V_--form of the wavefunction inside the actual rough surface by the V_+-form, then, the calculation becomes pretty straightforward and yields that the roughness induced diffuse scattering has the form (3.40) with

$$S_{\mathrm{dif}}^\varrho(Q') = \frac{V_{n0}^2 e^{-2M'_\varrho}}{|Q'_z|^2} \iint d^2 r_\parallel \Big(e^{-|q'_z|^2 g_\perp(r_\parallel)} - 1\Big) e^{iQ_\parallel \cdot r_\parallel} \tag{3.43}$$

with $2M_\varrho$ (2.42) being the roughness Debye–Waller factor and $g_\perp(r_\parallel) \equiv \langle |h(0)\, h(r_\parallel)|^2 \rangle$ the height-height correlation function which is conventionally approximated by a Gaussian distribution as long as $g_\perp(r_\parallel)$ does not diverge. Surface roughness occurs at different length scales, thus, rough surfaces should exhibit a "broken" self-similarity ("self-affinity"), where the z-direction does scale differently than the $r_\parallel$-direction (see [Mandelbrodt, 1982; Alexander, 1987])[10]. This can be approached by a correlation function of the form

$$g_\perp(r_\parallel) = 2\varrho^2 \left(1 - e^{-(r_\parallel/\xi)^h}\right) \tag{3.44}$$

with a correlation range ξ and the self-affine exponent $h \equiv (3 - d_h) \in [0, 1]$ which produces a "jagged" surface for small values and smooth hills upon approaching 1 ($2 \leq d_h \leq 3$ is the Hausdorff dimension of the surface). In Sect. 5.3 we will discuss $g_\perp(r_\parallel)$ close to the surface roughening temperature.

Sinha et al. [1988] have discussed the evanescent diffuse scattering associated with self-affine surface roughness. Figure 3.6b shows the expected diffuse scattering around the specular beam as a function of the incident angle α_i for a rough surface characterized by $\xi = 7000$ Å, $\varrho = 7$ Å and a self-affinity parameter $h = 0.2$. During the scan the angular difference $\alpha_f - \alpha_i$ is kept fixed at 1°. The diffuse intensity peaks at a specular condition $\alpha_i = \alpha_f$ (= 0.5° in this case) as expected from the scattering law (3.43) and in addition for $\alpha_i = \alpha_c$ and $\alpha_f = \alpha_c$ due to the action of the transmission functions occuring in (3.40). For grazing incidence and exit geometries Q'_z is generally so small that the exponential term in the integral may eventually be linearized. Then (3.43) gives rather simply the Fourier transform of the height-height correlation function. On the experimental side surface roughness-induced diffuse scattering has not been studied so far in a systematic way, some recent studies have been reported using x-rays. (It should be noted that the analogous theory applies for evanescent x-ray scattering.) The roughness of quartz-glass and SiC surfaces has been investigated in the soft x-ray regime ($\lambda = 100$ Å $- 25$ Å) by Birken et al. [1990]. In this case $\varrho \ll \lambda$ ("Rayleigh regime") and one finds that the vector perturbation theory developed by Elson [1984] gives a good agreement with the observed diffuse scattering distribution. A crucial test of the Sinha theory has been undertaken by Weber and Lengeler [1991] who measured the roughness-induced diffuse scattering from Al polycrystalline surfaces (Fig. 3.7). The experiments have been performed at the RÖMOI-station at HASYLAB using a wavelength of $\lambda = 1.77$ Å. The full line in Fig. 3.7 is a theoretical fit according to the theory by Sinha et al. [1988] and gives $\varrho \simeq 40$ Å and $\xi \simeq 4000$ Å. It turns out that the best agreement between the DWBA theory and the experimental data is obtained, when the roughness-enhanced transmission functions (2.19) as given by Nevot and Croce [1980] are used in (3.40) and when the full, i.e. non-linearized, structure factor (3.43) is employed.

---

[10] Conventional self-similarity would mean that a surface would look rough at any length scale, while at large scale any real surface looks smooth.

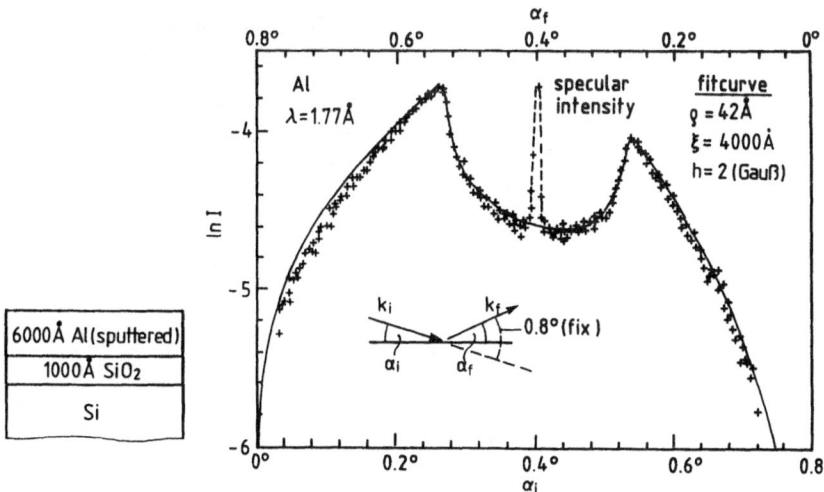

Fig. 3.7. Observed surface roughness-induced diffuse scattering from a polycrystalline Al surface [Weber and Lengeler, 1991] (the sample is shown on the left)

3.3.2 Magnetic Scattering

Mills [1975] presented a theory of magnetic surface scattering of low-energy electrons which was later extended to neutrons by Mazur and Mills [1982]. With the advent of first rigorous results on the critical behaviour of semi-infinite matter Dietrich and Wagner [1985] considered magnetic evanescent neutron scattering from this view point. Here I sketch the present stage of the theory.

At the surface of magnetic matter the B-field (3.8) has the form

$$
\begin{aligned}
B_{\mp}(r) = & \frac{\mu_0}{4\pi} \int_{V_+} d^3 r' \left(M(r') \cdot \nabla' \right) \nabla' \frac{1}{|r - r'|} \\
& + \begin{cases} 0 & \text{for } r \in V_- \\ \mu_0 M(r) & \text{for } r \in V_+ \end{cases} .
\end{aligned}
\tag{3.45}
$$

Since $B_- \neq 0$, although $M(r) = 0$ in V_-, the magnetic neutron scattering amplitude V_m^{fi} is

$$
\begin{aligned}
V_m^{\mathrm{fi}} &= \left\langle \Psi_f \left| -\mu_n \cdot B_{\mp} \right| \Psi_i \right\rangle = V_{m+}^{\mathrm{fi}} + V_{m-}^{\mathrm{fi}} \\
&= -\mu_n \cdot \left(\int_{V_+} d^3 r \, \Psi_f^* B_+(r) \Psi_i + \int_{V_-} d^3 r \, \Psi_f^* B_-(r) \Psi_i^* \right) ,
\end{aligned}
\tag{3.46}
$$

where the appropriate solutions (3.21) of the evanescent neutron wave functions in the two halfspaces $V_\pm$ have to be used as discussed before. Thus, we get the amplitude

$$
V_{m+}^{\mathrm{fi}} = -\mu_n \cdot \beta_+^{\mathrm{fi}} \int_{V_+} d^3 r \, B_+(r) e^{-iQ_\parallel \cdot r_\parallel} e^{-i(k_{fz}'^* - k_{iz}')z}
\tag{3.47}
$$

47

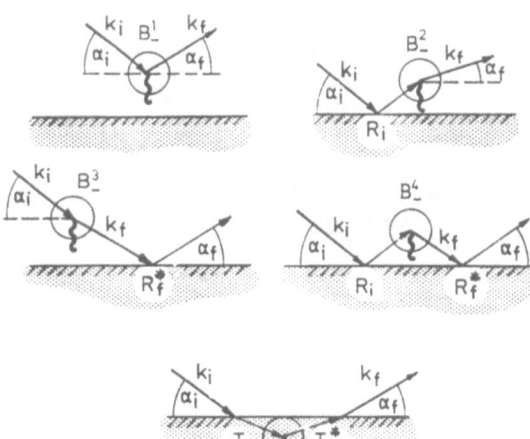

Fig. 3.8. Neutron diffuse scattering contributions to the forward intensity at a ferromagnetic interface: top and middle view show the contributions from the dipolar fields in the vacuum, the bottom view shows the near-surface evanescent diffuse scattering

which is the scalar product of μ_n and the Fourier–Laplace transform of the near-surface B-field, and in addition

$$
\begin{aligned}
V_{m-}^{fi} &= B_-^1 + B_-^2 + B_-^3 + B_-^4 \\
&= -\mu_n \cdot \int_{V_-} d^3r\, B_-(r)\, e^{-iQ_\| \cdot r_\|} \\
&\quad \times \left(e^{+iq-} + R_i e^{-iq+} + R_f^* e^{+iq+} + R_f^* R_i e^{-iq-} \right)
\end{aligned}
\tag{3.48}
$$

from the dipolar fields outside the sample which consists of four different contributions (Fig. 3.8). In (3.48) $q_\pm = k_{iz} \pm k_{fz}$ (in V_-). After performing one "half-space" integration we obtain V_m^{fi} [Mazur and Mills, 1982; Dietrich and Wagner, 1985]:

$$
\begin{aligned}
V_m^{fi} = -\mu_0 \mu_n \cdot \int_{V_+} d^3r \Big(&\beta_+^{fi} e^{iQ' \cdot r} Q' \times (M(r) \times Q') \\
&- \frac{1}{2Q_\|^2} \beta_-^{fi} e^{i\Omega' \cdot r} \Omega'(\Omega' \cdot M(r)) \Big)
\end{aligned}
\tag{3.49}
$$

with the Fresnel functions $\beta_+^{fi} = T_f^* T_i$, as before, and

$$
\beta_-^{fi} = Q_\| \left(\frac{1}{Q_\| + iq_-} + \frac{R_i}{Q_\| - iq_+} + \frac{R_f^*}{Q_\| + iq_+} + \frac{R_f^* R_i}{Q_\| - iq_-} - \frac{T_f^* T_i}{Q_\| + iQ_z'} \right)
\tag{3.50}
$$

and the complex space frequency

$$
\Omega' = Q_\| + iQ_\| z/z \quad (z \text{ in } V_+) \quad .
\tag{3.51}
$$

For a rigorous discussion of the apparently complicated form of (3.49) I refer the reader to the clearly arranged article by Dietrich and Wagner [1985]. Here I want to add some comments which may be helpful for the general reader:

The first term in (3.49) is the, say, "usual" term in evanescent scattering, namely the Fourier–Laplace transform of a microscopic quantity, here of $\mu_n \cdot P'_{e\perp}$ with

$$P'_{e\perp} = Q' \times (M \times Q')/Q'^2 \quad . \tag{3.52}$$

$\mathrm{Re}\{P'_{e\perp}\}$ is, as in bulk scattering, the magnetization density component normal to the momentum transfer. The second term accounts for scattering from the dipolar fields in V_- which appears at the surface of ferromagnets. The complicated nature of this scattering process is reflected in the uncomfortable form of β^{fi}_- (3.50) and in the appearance of the peculiar momentum Ω' (3.51) which induces an exponential damping of the second term in (3.49) in the z-direction. This damping, however, does not depend on Q'_z, but on the parallel momentum transfer[11] $Q_\parallel$, therefore, this term allows bulk scattering contributions to enter (3.49). Pure neutron scattering signals from surface-related ferromagnetism are only obtained when this bulk scattering is avoided: In the scattering geometry of Fig. 2.5, where both angles α_i and α_f are $O(\alpha_c)$, it is quite easy to show that β^{fi}_- is only nonzero in the small angle scattering regime ($Q_\parallel \to 0$). Around this forward scattering regime the second term in (3.49) exhibits a diverging behaviour which has not yet been analyzed rigorously. When both angles α_i and α_f are far beyond α_c, (3.49) turns into the associated bulk form[12], where again $\beta^{fi}_- = 0$ and, of course, $Q' = Q$.

Evanescent neutron *scattering* will have promising and unrivaled applications in interfacial magnetism [Ankner et al., 1991] and in particular in surface antiferromagnetism which is not detectable in the specular beam. In this case $B_- = 0$ and all the complications discussed above disappear. One example of some intriguing problems of current interest which can be investigated by the scattering of evanescent neutron waves is the magnetic long range order at the interface between a ferromagnet and an antiferromagnet (e.g. the Ni-NiO interface or the Fe-Cr interface [Saurenbach et al. [1988]).

---

[11] The integrations $\int d^3 V_\pm$ are performed as surface integrals by means of the second Green formula. This finally leads to an integral of the form

$$\int_0^\infty r'\, dr'\, J_0(Q_\parallel r')/\sqrt{r'^2 + z^2}$$

which results in a $Q_\parallel$-dependent exponential decay (see B.2) in Appendix B.

[12] It was remarked by Dietrich and Wagner [1985] that the expression proposed by Mazur and Mills [1982] does not behave this way.

3.4 Scattering Experiments with Evanescent Neutrons

While the study of the neutron specular intensity profile is a well established tool in the surface science, in particular applied to surface magnetism [Felcher, 1981; Felcher et al. 1984; Pescia et al. 1987], liquid surfaces [Hayter et al. 1981; Penfold and Thomas, 1990] and polymer films [Russell et al. 1989; Stamm et al. 1989] neutron scattering under the condition of total external reflection is still in an early stage. This is owing to the very weak surface scattering cross section and the extreme grazing angle scattering geometry which both ask for high brilliance in the incident beam and new neutron instrumentation. Hence this field has been almost exclusively attacked with x-rays provided by synchrotron sources, and only a few attempts to render neutron scattering surface sensitive by grazing angle geometries have been undertaken. In the following I will give a discussion of these experiments reported in literature. Before that I also mention a completely different approach to render neutrons surface sensitive by Al Usta et al. [1990] who were able to detect for the first time socalled *"neutron truncation rod scattering"* from the surface of perfect crystals. In this study the authors exploited the fact that the crystal truncation at the surface can be detected as tiny diffuse intensity wings which extend from the Bragg intensity normal to the surface [Andrews and Cowley, 1985; Robinson, 1986], thereby following the law

$$I_{\text{TR}}(q_z) \propto q_z^{-2} \tag{3.53}$$

with $q_z \equiv (G_{hkl} - Q)_z$ (see Fig. 3.9b). The measurements have been performed at the special instrument S21 (Fig. 3.9a) at the high flux reactor of the Institut Laue–Langevin in Grenoble. In order to achieve the necessary resolution a perfect crystal monochromator and analyzer have been used. The scattering geometry in the reciprocal space is illustrated in Fig. 3.9b: G_{hkl} denotes the reciprocal lattice vector associated with the Bragg planes d_{hkl}. The resolution function can be visualized as a "needle" with $\Delta q_x = 8.8 \times 10^{-4} \text{ Å}^{-1}$ and $\Delta q_z = 6.0 \times 10^{-4} \text{ Å}^{-1}$ in the two directions q_x and q_z which are related with the sample rotation $\Delta\omega_S$ and analyzer rotation $\Delta\omega_A$ by

$$q_x = \frac{2\pi}{\lambda} \left(2\Delta\omega_S - \Delta\omega_A\right) \sin\theta \tag{3.54}$$

$$q_z = \frac{2\pi}{\lambda} \Delta\omega_A \cos\theta \quad .$$

The location of the neutron truncation rod (given by $q_x = 0$, or $2\Delta\omega_S = \Delta\omega_A$) is indicated in Fig. 3.9b as "TR", while the dashed lines denoted "M" and "A" show the location of "spurious" diffuse scattering from the dynamical wings of the monochromator and analyzer reflection, respectively. The intensity "M" is Bragg reflected from the sample and the analyzer crystal, thus appears, whenever $\Delta\omega_S = \Delta\omega_A$ holds. Analogously, the intensity "A" appears for $\Delta\omega_S = \Delta\omega_M =$

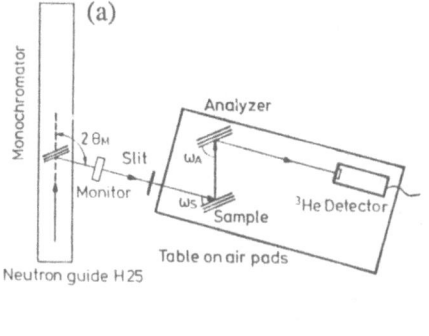

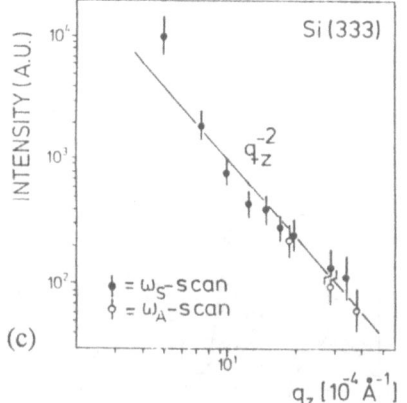

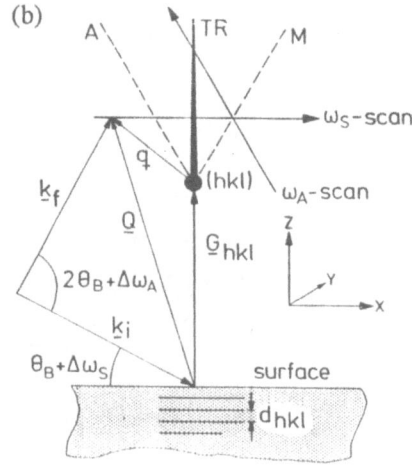

Fig. 3.9. Measurement of neutron truncation rod scattering [Al Usta et al., 1990] (a) schematic view of experimental arrangement using the special instrument S21; (b) scattering geometry in reciprocal space and scattering contributions in the three-crystal-mode: M = monochromator wing, A = analyzer-wing, TR = truncation rod (see text); (c) surface scattering intensity versus q_z

0 (since the monochromator remains fixed). In the experiment two different scan modes have been performed: a rotation of the sample for a fixed detuning of the analyzer ("ω_S-scan" in Fig. 3.9b) and a rotation of the analyzer for a fixed setting of the sample rotation ("ω_A-scan" in Fig. 3.9b). Figure 3.10 shows typical examples of the observed intensity distributions along ω_s-scans for various detunings $\Delta\omega_A$ from the ideal Bragg position. One finds the expected three-peak structure with the intensities denoted M and A as the spurious diffuse scattering from the monochromator and analyzer crystals as discussed above and the distinct scattering denoted TR which results from the mere presence of the sample surface. According to Fig. 3.9c these intensities (hatched areas in Fig. 3.10) follow indeed the predicted powerlaw decay (3.53). Note here the very low counting rate of the surface signals. Since the momentum transfer in such surface experiments is purely normal to the surface, neutron truncation rod scattering may be performed complementary to grazing angle scattering experiments which will be described in the following.

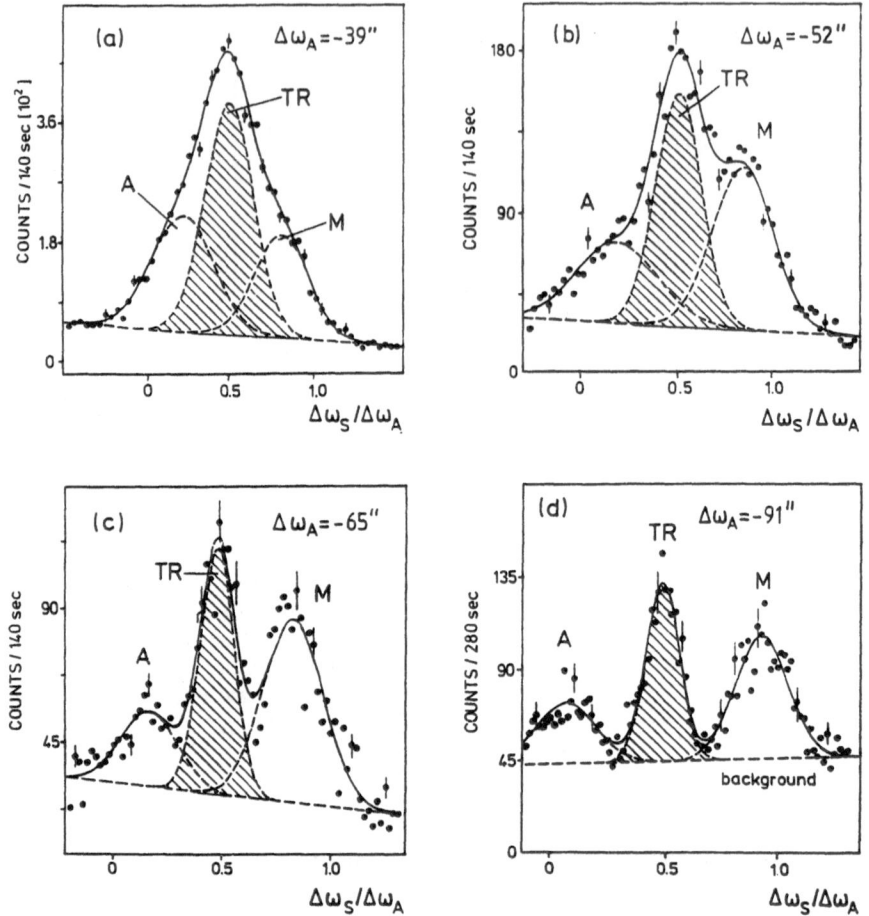

Fig. 3.10a–d. Diffuse neutron scattering around the Si(333) reflection as observed along ω_S-scans for various setting of ω_A [Al Usta et al., 1990] (see text)

3.4.1 Asymmetric Grazing Incidence Diffraction

A detailed experimental and theoretical study of neutron scattering at grazing incidence has been reported by Zeilinger and Beatty [1983] who investigated the $(0\bar{2}2)$ Bragg scattering from a Si(211) surface in the strongly asymmetric scattering geometry, where the incidence angle α_i with respect to the surface is forced to $O(\alpha_c)$ by tuning the neutron wavelength to (see also Dosch [1987])

$$\lambda_\phi = 2d_{hkl}\sin\phi \quad , \tag{3.55}$$

where ϕ is the angle between the Bragg planes (d_{hkl}) and the surface (Fig.3.11a). As a consequence, k_f emerges from the surface at an angle close to 2ϕ. In the experiment the $(0\bar{2}2)$ Bragg reflection was studied as a function of α_i with $\lambda_\phi = 3.137$ Å ($\phi = 54.74°$) resulting in a critical angle of $\alpha_c = 2.5$ mrad. Figure 3.11b shows the obtained peculiar intensity distribution before and after

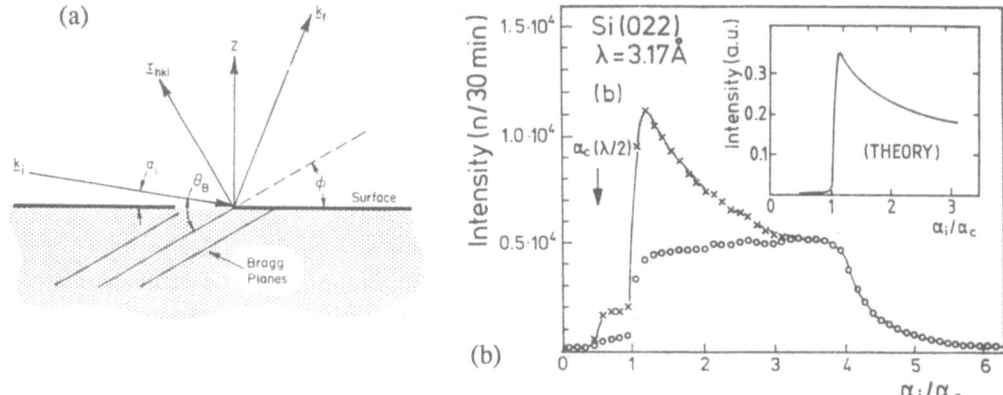

Fig. 3.11a,b. Asymmetric grazing angle neutron scattering: (a) scattering geometry (after Dosch [1987]); (b) observed scattering intensity at the Si(211) surface [Zeilinger and Beatty, 1983]

correcting the data for the α_i-dependent *"Fankuchen"*-broadening of the beam cross section (see Fankuchen [1938]).

Since we are dealing with the surface of a perfect crystal, the understanding of the details of this asymmetric Bragg profile requires, as in the bulk, a dynamical theory of neutron scattering which goes beyond the first Born approximation (3.40). Conveniently in treating dynamical neutron scattering effects one considers the Schrödinger equation (3.1) in its Fourier transformed form (see [Batterman and Cole, 1964])

$$\left(-\frac{\hbar^2}{2m}(Q+G)^2 - E\right)\Psi(G) = -\sum_{G'} V(G-G')\Psi(G') \quad , \qquad (3.56)$$

where G and G' are reciprocal lattice vectors of the solid. When one Bragg node G^* is excited as in our case, the sum $\sum_{G'}$ goes over $G' = 0$, G^* (2-beam case). Of course, the Schrödinger equation (3.1) and (3.56) contains evanescent neutron waves, therefore the dynamical scattering theory does not need to be extended to the case of grazing angle diffraction, even though the proper calculation of evanescent Bragg scattering may be slightly more complicated than the conventional bulk case. The grazing angle theory for the x-ray case has already been treated in the early paper by Farwig and Schürmann [1967]. Meanwhile many theoretical studies of the application of the dynamical scattering theory to the case of total external reflection condition have been published which apply both for neutrons and x-rays. We refer the reader to the work by Rustichelli [1975], Afanase'v and Melkonyan [1983], Cowan [1985] and Bernhard et al.[1987]. The inset in Fig. 3.11b shows the result of a numerical calculation based on (3.56) which reproduces the salient features of the experimental curve, in particular the asymmetric line shape and the position of the Bragg maximum which is attained for $\alpha_i \geq \alpha_c$. The latter is, according to our experience gained from DWBA, partly owing to the neutron transmission of the interface which is here implicitly embedded in the dynamical equations. Zeilinger and Beatty also investigated how the angular position of the Bragg peak is altered by the strong grazing angle refraction effects from the position deduced from conventional bulk dynamical theory. The observed refraction induced deviations agree well with the analytical expression given by Rustichelli [1975].

3.4.2 Grazing Angle Diffraction

Neutron scattering under grazing incidence from Bragg planes perpendicular to the surface has been reported by Ankner et al. [1989] who measured the (111) Bragg scattering of a Si($1\bar{1}0$) surface with a neutron wavelength λ = 4.43 Å. While a certain exit angle range was integrated over the detector aperture, the Bragg scattering was recorded as a function of the incidence angle between $\alpha_i = 0.5\,\alpha_c$ and $\alpha_i = 2.0\,\alpha_c$ and was found to exhibit a maximum at $\alpha_i/\alpha_c \simeq 1$. Unfortunately the published experimental evidence is very little and not described in detail. Since in addition the presented theoretical curves show a rather unclear behaviour, it is somewhat questionable to the author as to how much a surface signal is really contained in the observed scattering.

It was pointed out by Al Usta et al. [1991] that surface sensitive neutron scattering is only possible by the meticulous control of *both* the grazing incidence and exit angle, mainly for two reasons:

a) The generally very low absorption cross section of neutrons with matter allows transmitted bulk scattering to contribute to the signal which, however, can be distinguished from the surface signal by the missing refraction effect in the exit angle (see below),

b) the surface sensitivity can significantly be enhanced when two grazing angles occur.

For α_i- and α_f-resolved ("depth controlled") grazing angle Bragg scattering experiments a conceptionally new instrument (EVA *nescent wave diffractometer*) has been realized at the high flux reactor of the Institut Laue–Langevin, Grenoble (see Al Usta et al. [1991]). A side view of the instrument and a close-up up of the sample mounting and positioning device is shown in Fig. 3.12. The slit system $C_1 - C_2$ taylors the incident neutron beam to a useful vertical size of $\sigma_{iv} = 0.1$ mm and a necessary vertical divergence of σ_{iv}^l(FWHM) = 0.35 mrad (note that the neutron guide tube supplies a neutron beam with $\Delta z_0 = 120$ mm and $\alpha_0 = 20$mrad). The instrument is supplied with moderately cold neutrons of wavelength $\lambda = 5.5$ Å provided by a PG(002) monochromator (Bragg angle $\theta_M = 60°$) with an intrinsic mosaicity of $\eta = 30$ mrad. Thus, the wavelength uncertainty of the neutron beam at the sample (with typical linear dimensions $d_S \simeq 40$ mm) is given by [Dorner, 1972; Dorner and Comes, 1977]

$$\Delta\lambda/\lambda = \left(\frac{(\alpha_0\alpha_1)^2 + (\alpha_0\eta)^2 + (\alpha_1\eta)^2}{\alpha_0^2 + \alpha_1^2 + 4\eta^2} \right)^{1/2} \cot\theta_M \qquad (3.57)$$

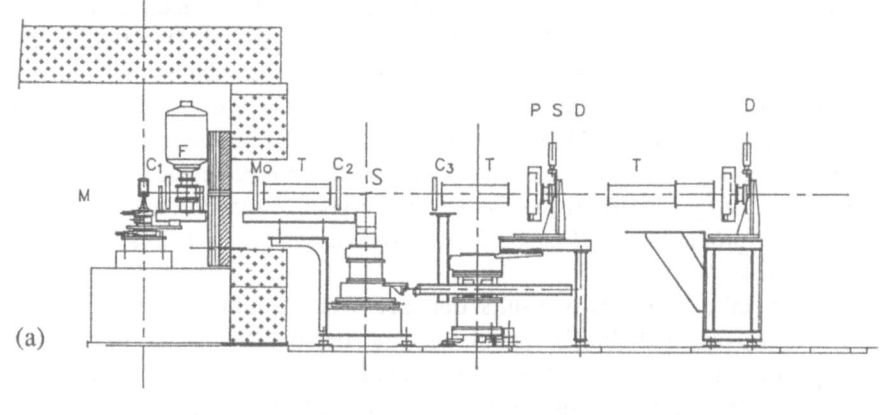

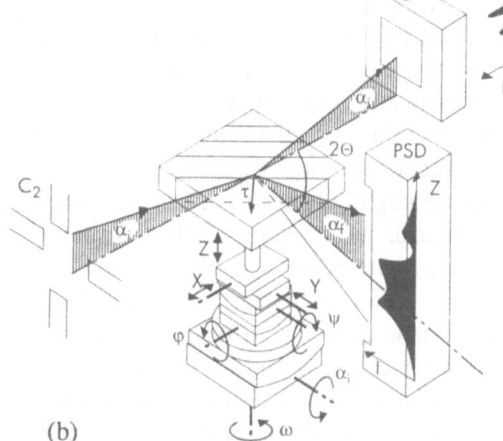

Fig. 3.12. Evanescent neutron wave diffractometer EVA [Al Usta et al., 1991] **(a)** side view upstream of the neutron guide: M = monochromator, F = Be-filter, $C_1 - C_2$ = vertical and horizontal collimation slits, C_3 = horizontal slits, T = vacuum tubes, PSD = position sensitive detector, D = specular beam detector **(b)** close-up of sample manipulation stage: sample surface alignment via $x - y - z$ translation stage and ψ- and φ-tilt stage; sample scan via ω-rotation stage and α_i-stage

with $\alpha_1 \equiv 2d_S/L_{MS}$ ($L_{MS} \simeq 2m$ is the distance monochromator-sample giving $\alpha_1 \simeq 40$ mrad), or $\Delta\lambda/\lambda = 0.009$. From geometric considerations one finds that the neutron intensity at the sample position is

$$I_0 = R_{PG} \frac{\delta\phi}{\delta\lambda} \Delta\lambda \frac{\beta_0 \sigma'_{iv}}{\beta_v^2} \quad , \tag{3.58}$$

where $R_{PG} \simeq 0.85$ is the reflectivity of the PG(002) crystal, $\delta\phi/\delta\lambda$ is the spectral neutron flux of the neutron guide and

$$\beta_v = \left[\left(2\eta \sin\theta_M\right)^2 + \beta_0^2 \right]^{1/2} \quad . \tag{3.59}$$

β_0 is the vertical divergence of the neutron guide. Taking $\delta\phi/\delta\lambda = 8 \times 10^8$ n/(cm^{2}sÅ) and $\beta_0 \simeq \alpha_0$ one finds roughly $I_0 \simeq 1 \times 10^6$ n/(cm^{2}s). The

55

actually observed neutron intensity at the sample with a neutron beam of σ_{iH}(FWHM) = 4 mm, σ_{iV}(FWHM) = 0.2 mm and σ'_{iV}(FWHM) = 0.4 mrad is around 2×10^5 n/(cm^{2}s). We will see in what follows that the evanescent neu-neutron Bragg scattering cross section is so small that one observes only a few counts per second in the PSD. Since this poses a serious problem in such experiments one would want to increase I_0. A current effort is to replace the PG(002) monochromator by a socalled "Si-Ge gradient crystal"-monochromator [Magerl, 1990] which produces $\Delta\lambda/\lambda \simeq 0.02$. This would lead to an increase of I_0 by roughly a factor 4 to 5.

By way of example two results obtained from a InP(1$\bar{1}$0) surface and a CaF$_2$($\bar{2}$11) surface will be discussed in some detail. The two systems differ strongly in the neutron absorption cross section, InP exhibits strong absorption leading to a 1/e-absorption length $l_a = 0.9$mm, while CaF$_2$ has $l_a = 31$mm. Figure 3.13a shows the grazing angle CaF$_2$(111) Bragg intensity versus the exit angle for a fixed incidence angle α_i close to α_c. One finds three different contributions to the Bragg intensity which are scattered at different exit angles: The peak at $\alpha_f/\alpha_c = 1$ (denoted "S" in Fig. 3.13a) is the evanescent Bragg scattering as in the x-ray case. At $\alpha_f/\alpha_c \simeq 0$ the transmitted diffracted intensity ("T") can be detected which originates from neutrons impinging onto the sample surface and, after Bragg reflection, re-escaping into the vacuum through the sample edge. This effect which is only observable because of the negligible absorption demonstrates very nicely the existence of evanescent neutron wave fields inside the sample which travel parallel to the surface and thus emerge at $\alpha_f = 0$. However, for the application to surface science one has to note that the surface sensitivity of this beam is only moderate. The intensity denoted "B", on the other hand, is a spurious bulk scattering due to neutrons which enter and leave the medium through the side edges and thus suffer no refraction effect at all. Consequently, this bulk Bragg intensity is found at $\alpha_f = -\alpha_i$. Contaminations of such spurious bulk scattering are usually inevitable, when the exit angle of the scattering is not

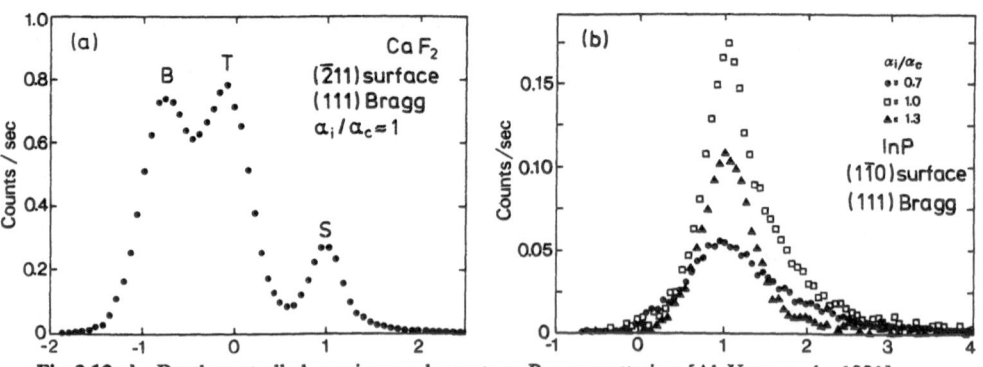

Fig. 3.13a,b. Depth-controlled grazing angle neutron Bragg scattering [Al Usta et al., 1991]. α_f-profiles of the (111) Bragg reflection at: (a) a CaF$_2$(211) surface; (b) an InP(110) surface

carefully analyzed. In InP the absorption cross section is large enough to wipe out the "T" and the spurious "B" beam, and only the most surface sensitive Bragg intensity is left as shown in Fig. 3.13b for various incidence angles.

From this scattering study it becomes quite apparent that a proper grazing angle neutron scattering experiment requires a meticulous analysis of the nature of the various Bragg scattering processes in order to identify the correct surface scattering signal. In the above discussion we tacitly assumed that the associated Bragg planes were ideally normal to the surface. In general, however, the situation is even more complex, because one has to account for a miscut $\Delta\phi$ which may produce an angular mixing of the three beams (B, T and S). The only way of properly disentangling the various contributions is the measurement of the different refraction effects which suffer the various Bragg beams: The B-beam suffers no refraction at all, while in the T-beam k_i gets refracted and in the S-beam both k_i and k_f suffer refraction. This leads to α_f - α_i-relations which are characteristic finger prints for the various beams (Fig. 3.14). This first systematic experimental study of grazing angle neutron scattering made also clear that the applicability of this technique is severely limited by the very low scattering intensity. Typical counting rates in one element of the position sensitive detector are as low as 0.2 cps (!) for evanescent Bragg scattering. So the conclusion has to be that on the one hand grazing angle neutron Bragg scattering is feasible, though it requires a careful analysis of the k_f-distribution, that on the other hand any measurement of diffuse neutron scattering under grazing angles will face a tremendous intensity problem.

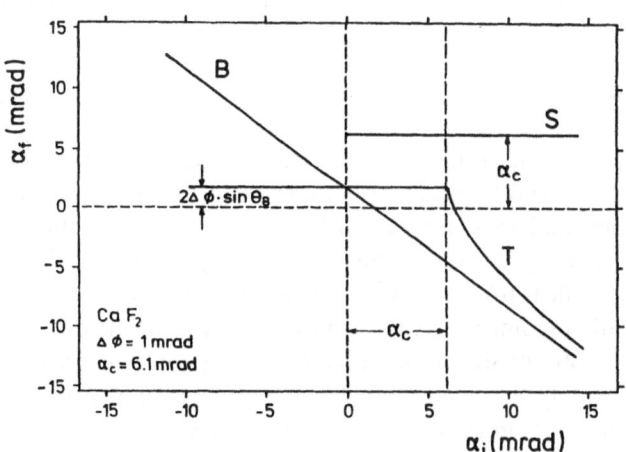

Fig. 3.14. Refraction effect in neutron Bragg scattering (schematic). α_f position of bulk intensity (B), transmitted intensity (T) and surface intensity (S) versus α_i

4. Semi-Infinite Critical Systems

Since the observation of critical opalescence in CO_2 bulk-critical phenomena have been studied to a great extent. The theoretical efforts to understand criticality started with van der Waals. Later Landau pioneered the mean field theory of critical phenomena which could provide a first insight into the universality of critical behaviour, however, failed to produce the observed critical exponents. This breakdown of the "classical" theory near the critical point led in a phenomenological way to the concept of scaling. The physical origine of scaling was provided by the renormalization group (RG) theory, its fruits are scaling relations between critical exponents and precise quantitative predictions for the values of the critical exponents. Many sophisticated experiments have been undertaken in the last decades to measure bulk critical behaviour with high precision in order to test these predictions, and today one can safely say that the RG theory of bulk-critical phenomena has fully been confirmed by the entire body of experimental work.

The fundamental mechanism behind the universal behaviour of critical phenomena is the unlimited growth of the system-specific correlation length ξ upon approaching T_c,

$$\xi = \xi_0^{\pm} |t|^{-\nu} \quad (t \to 0_{\pm}) \quad , \tag{4.1}$$

where $t \equiv (T - T_c)/T_c$ is the reduced temperature and $\nu \simeq 0.67$ ($\simeq 2/3$) a universal exponent. Since the real world is necessarily finite-sized, the understanding of critical phenomena is incomplete as long as surface effects are not included. This led Fisher and Barber [1972] to introduce the concept of *"finite-size scaling"* (nowadays also of great practical use as in Monte–Carlo simulations of critical behaviour in order to handle the impetuous growth of ξ). However, the detailed knowledge of the rounding-offs of the critical divergences by the finite size of the system does not at all imply that the critical behaviour close to a surface is already understood. The appropriate approach to attack this task is the study of semi-infinite systems which fill up one halfspace and provide both the presence of a surface and ample space to allow ξ to diverge. The investigation of such *"semi-infinite criticalities"* was sparked by Fisher [1971] and Binder and Hohenberg [1972] who later presented a pioneering Monte–Carlo study [Binder and Hohenberg, 1974]. Meanwhile the semi-infinite mean-field theories (see [Binder, 1983]) and the semi-infinite RG theories (see [Diehl, 1983; Wagner, 1985]) are essentially worked out and provide the experimentalist with precise predictions concerning the surface-modified critical behaviour. In the following I will discuss some of these predictions along with scattering experiments from solid surfaces which allow first critical tests of the theory.

Regard the surface of a semi-infinite system (Fig. 4.1a). Atoms close to a surface have less neighbours and are in general less tightly bound than their bulk counterparts, unless the remaining surface couplings (J_1) are so much enhanced that they balance or even overcompensate the reduced coordination number. In addition to this surface enhancement c,

$$c \equiv 1 - 4[(J_1/J) - 1] \qquad (4.2)$$

the coupling of the surface atoms to an eventually present surface field h_1 has to be taken into account. It turns out that c and h_1 are the two relevant scaling fields which govern the critical behaviour of semi-infinite systems [Binder, 1983]. For $h = h_1 = 0$ (h is a bulk field) several new universality classes emerge depending on the action of c (Fig. 4.1b): "O" denotes the *ordinary transition*, where the order at the surface occurs at the same temperature as in the bulk, "S" denotes the *surface transition* with a surface order above the bulk critical temperature. This transition and the *extraordinary transition* "E" occur for a sufficiently strong surface enhancement c. The ordinary and the extraordinary transition are separated by the *special transition* "SB".

I want to emphasize at this point two nontrivial results of the theory:

a) The truncated translational invariance at the surface of semi-infinite systems does not evoke a second critical length scale, thus, the surface-modified critical long range order (LRO) and its critical fluctuations are governed by the bulk correlation length ξ (4.1).

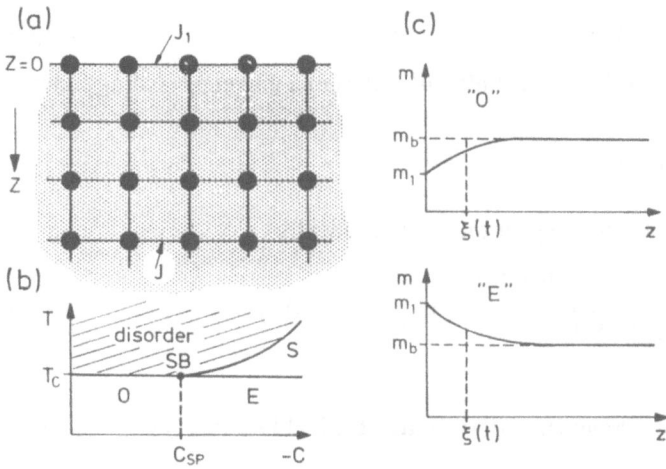

Fig. 4.1a-c. Semi-infinite solid. (a) Ising model with bulk coupling J and surface coupling J_1; (b) general phase diagram: c = surface enhancement (4.2), O = ordinary transition, E = extraordinary transition, S = surface transition, SB = special transition; (c) order parameter profiles $m(z)$ associated with the ordinary and extraordinary transition

59

b) As a consequence there exist new scaling relations which connect surface-critical and bulk-critical exponents.

The experimental challenges are the reliable measurement of the surface-critical exponents and the critical experimental test of the new scaling relations.

4.1 Order Parameter Near a Free Surface

The central quantity in the description of phase transitions is the associated order parameter (OP) $m \propto \delta f/\delta h$ (f = singular part of free energy; h = conjugate field) which measures the degree of LRO in the system. Due to the truncation of translational invariance at the surface the OP below the critical temperature becomes necessarily inhomogeneous along the z_+-axis. Intuitively one would assume a near-surface OP profile as shown in Fig. 4.1c which is associated with the ordinary transition. Also shown is the OP profile associated with an extraordinary transition "E", where the bulk orders in the presence of an already ordered surface.

In the following we consider the ordinary transition: For $h = h_1 = 0$ the OP $m_1(t) \equiv m(z = 0, t)$ at the surface of a semi-infinite system vanishes as

$$m_1(t) \propto |t|^{\beta_1} \quad (t \to 0_-) \tag{4.3}$$

with reduced temperature $t \equiv 1 - T/T_c$, where $\beta_1 \simeq 0.8$ [Diehl and Dietrich, 1981; Ohno et al., 1984; Binder and Landau, 1984] is a universal surface exponent. From RG theory one knows that the asymptotic OP profile has scaling form [Diehl and Dietrich, 1981],

$$m(z, t, h, h_1) = m_b \, f\left(z/\xi, \, h/|t|^{\Delta}, \, h_1/|t|^{\Delta_1}\right) \tag{4.4}$$

with $m_b \propto |t|^{\beta}$ ($\beta \simeq 0.3$) being the bulk OP and $f(x)$ a universal scaling function with the property

$$f(x, 0, 0) = f_0 x^{(\beta_1 - \beta)/\nu} \quad . \tag{4.5}$$

Thus, for any arbitrary z we expect that asymptotically

$$m(z, t) \propto |t|^{\beta} \left(\frac{z}{\xi_0^- |t|^{-\nu}}\right)^{(\beta_1 - \beta)/\nu} \propto |t|^{\beta_1} \quad (t \to 0_-) \quad . \tag{4.6}$$

4.1.1 Ferromagnetism Near the Ni(100) and EuS(111) Surfaces

Since low-energy electrons interact dominantly with the atoms of the toplayer, LEED and SPLEED measurements should be most sensitive to m_1 (4.3). The quantity which is actually observed with SPLEED is the scattering asymmetry between the scattering with the electron spin parallel and antiparallel to the surface magnetization. This signal turns out to be proportional to m_1 [Feder and Pleyer, 1982] and free from critical diffuse scattering contributions. The latter property allows one to measure the average surface magnetization closest to T_c.

So far two independent SPLEED experiments have been performed in order to measure the surface magnetization of the Heisenberg ferromagnet Ni close to the bulk Curie temperature ($T_c \simeq 630$ K), one by Celotta et al. [1979] at the Ni(110) surface, the other by Alvarado et al. [1982] at the Ni(100) surface. Celotta et al. [1979] showed that the m_1 vanishes approximately linearly with reduced temperature (thus, $\beta_1 \simeq 1$) which is distinctly different from the bulk value $\beta = 0.33$ and apparently close to the mean field surface value $\beta_1 = 1.0$. However, the experimental temperature range was only $0.5 > t > 0.2$, therefore it is not a priori clear, whether they have observed critical behaviour. In the Ni(100) experiment (Fig.4.2) a much wider temperature range was accessible. The surface magnetization decays according to a powerlaw with a critical exponent

$$\beta_1(\text{Ni}) = 0.825 \pm {}^{0.025}_{0.040} \quad . \tag{4.7}$$

The observed value of β_1 and the experimental fact that m_1 disappears at T_c (bulk) give a strong evidence that the Ni(110) surface belongs to the "ordinary" universality class.

A similar SPLEED study has been reported on the EuS(111) surface [Dauth et al., 1987]. EuS can be called a textbook Heisenberg ferromagnet with $T_c \simeq 17$ K. The experiments show that again m_1 drops to zero at T_c indicating an ordinary behaviour, however, one finds that

$$\beta_1(\text{EuS}) = 0.72 \pm 0.03 \tag{4.8}$$

which is somewhat less than the best theoretical value for this universality class,

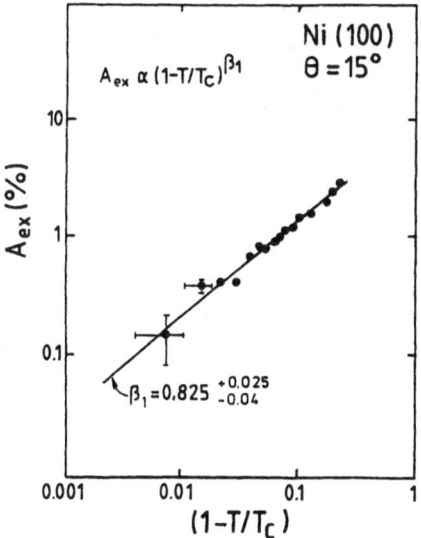

Fig. 4.2. Surface magnetization at the Ni(100) surface near T_c [Alvarado et al., 1982]

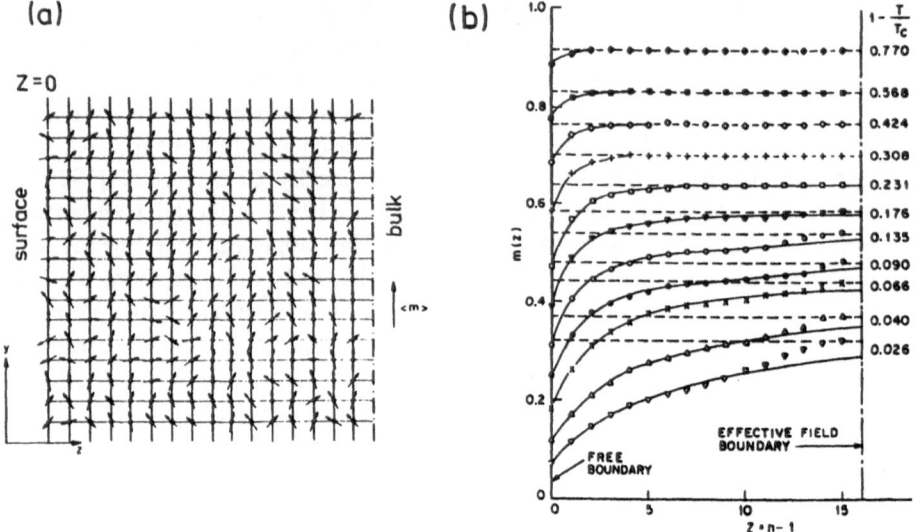

Fig. 4.3a,b. Monte–Carlo simulation of a Heisenberg ferromagnet [Binder and Hohenberg, 1974]. (a) Near-surface spin configuration at $t = 0.23$; (b) near-surface depth profiles of the laterally averaged magnetization at various temperatures

$\beta_1 = 0.84$ [Diehl and Nuesser, 1986]. One way out of this discrepancy is to assume that surface induces an anisotropy which drives the near-surface criticality more to an Ising-like behaviour.

The ordinary phase transition which has been observed in both cases is accompnied by an OP profile as in Fig. 4.1c. Binder and Hohenberg [1974] have calculated such OP profiles close to the bulk Curie temperature of a Heisenberg system (Fig. 4.3). It remains an intriguing experimental challenge to measure such critical magnetization profiles as a function of the temperature, a task which could be attacked by specular reflection of polarized neutrons. Similar to the experiment on a Ni surface (see Sect. 3.2.1) the flipping ratio of the specular intensity (Fig. 3.4b) could provide details of the critical tail of the near surface magnetization. Neutron experiments of that kind are currently in preparation.

4.1.2 Critical LRO at the Fe$_3$Al Surface

In the following we consider the surface of a binary alloy (AB) which undergoes a continuous order-disorder transition. In the (partially) ordered alloy one can distinguish two different site (α- and β-sites) which occur with the fractions y_α (y_β). When we call c_A (c_B) the actual concentration of the A- (B-) atoms and r_α (r_β) the fraction of α- (β-) sites which are occupied by the right atom A (B), then one may define an OP

$$m(T) \equiv \frac{r_\alpha(T) - c_A}{1 - y_\alpha} \tag{4.9}$$

which is 1 in the ideally ordered state ($c_A = y_\alpha$ and $r_\alpha = 1$) and 0 in the completely disordered state ($r_\alpha = c_A$). Note that the temperature dependence of $m(T)$ originates solely from $r_\alpha(T)$ ("*ordering density*").

When x-rays or neutrons undergo (bulk-) Bragg scattering from the ordered alloy two qualitatively different Bragg spots occur, socalled "*fundamental* (F)*"* reflections which originate from the average underlying lattice structure, and socalled "*superlattice* (SL)*"* reflections which are sensitive to the LRO $m(T)$. The associated Bragg amplitudes are (see e.g. [Warren, 1969])

$$F_F = N_{uc}(c_A f_A + c_B f_B)$$
$$F_{SL}(T) = m(T)(f_B - f_A)$$
(4.10)

with N_{uc} as the number of atoms in the unit cell and f_A (f_B) the form factor of atom A (B). In the neutron case f_A (f_B) is replaced by the associated coherent scattering length b_{cA} (b_{cB}). Thus, the precise measurement of $|F_{SL}(T)|^2$ allows one to obtain $m^2(T)$ in very direct way. Binary alloys which undergo continuous order-disorder phase transitions are not too numerous, the most convenient ones are β-CuZn (see [Als–Nielsen and Dietrich, 1967]), FeCo and Fe$_3$Al (see [Allen and Cahn, 1976]). Since $F_{SL} \propto (f_B - f_A)$, the most suitable system for x-rays is Fe$_3$Al which exhibits a line of continuous order-disorder phase transitions in the concentration range $c_{Al} = 0.25 - 0.31$ (Fig. 4.4a) at critical temperatures T_c around 500 K (depending on the actual composition). The phase diagram and the various FeAl structures have been studied by many authors [Lawley and Cahn, 1961; Epperson and Spruiell, 1969; Allen and Cahn, 1976].

Below T_c the alloy exhibits a DO_3 structure ("BiF$_3$ *structure*"), in the high temperature phase a $B2$ structure ("CsCl *structure*"). The higher-order transition $B2 \rightarrow DO_3$ eliminates the symmetry element along the $(100)a_0/2$, where $a_0 = 5.78$ Å denotes the lattice constant of the DO_3 supercell. For the sake of clarity I indicated in Fig. 4.4b how the lattice occupation changes in the CsCl unit cell during the order-disorder transition. The associated x-ray structure factors F_{hkl} can be calculated in the standard way, in particular, for (hkl) *unmixed odd* one finds that[1]

$$F_{hkl} = \begin{cases} \propto (f_{Fe} - f_{Al}) & \text{for } DO_3 \\ 0 & \text{for } B2 \end{cases},$$
(4.11)

thus, F_{hkl} (4.11) is a suitable superlattice reflection to monitor the degree of long range DO_3 order in the alloy during the phase transition. E.g. the scattering intensity around the (111) reciprocal lattice vector contains the information on the relevant OP and its local fluctuations.

Let us turn now to LRO at the border of such an alloy. Quite generally near-surface LRO is related to evanescent x-ray and neutron Bragg scattering by a complex Laplace transform (2.37), i.e. the evanescent superlattice amplitude is given by

---

[1] hkl notation associated with the supercell

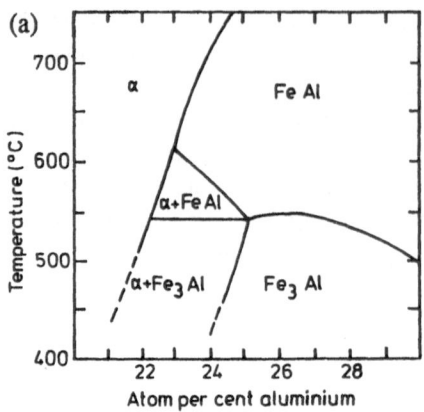

Fig. 4.4. The binary alloy FeAl; (a) part of the phase diagram [Allen and Cahn, 1976]; (b) high and low temperature phases of Fe_3Al in the CsCl subcell

(a)

Temperature (°C)

700

α   Fe Al

600

α+Fe Al

500

α+Fe₃ Al   Fe₃ Al

400

22   24   26   28

Atom per cent aluminium

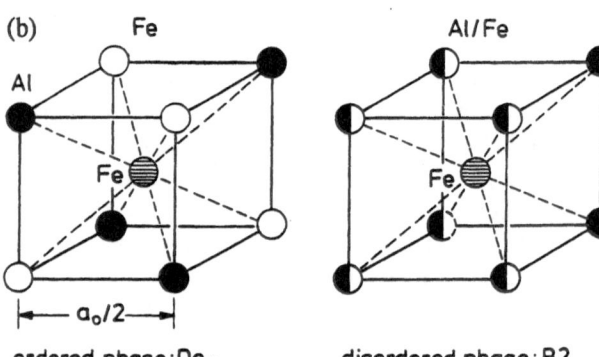

(b)   Fe   Al/Fe

Al

Fe   Fe

$a_0/2$

ordered phase:Do_3 disordered phase: B2

$$F_{\mathrm{SL}}(Q'_z, t) \propto (f_B - f_A) \int_0^\infty m(z,t) e^{iQ'_z z} dz \quad . \tag{4.12}$$

Here the critical OP profile $m(z,t)$ is given by (4.6), thus, the Bragg intensity assumes asymptotically for $t \to 0_-$ the scaling form [Dietrich and Wagner, 1984] (see Appendix B.3–4)

$$I_B(Q'_z, t)\big|_{t \to 0_-} \propto |t|^{2\beta}(1/\xi)^{2(\beta_1-\beta)/\nu} \left| \int_0^\infty z^{(\beta_1-\beta)/\nu} e^{iQ'_z z} dz \right|^2$$

$$\to |t|^{2\beta_1} \left(-iQ'_z \xi_0^-\right)^{2(\beta-\beta_1-\nu)/\nu} \quad . \tag{4.13}$$

Note that $I_B(Q'_z, t)$ is the average Bragg intensity within a surface layer of thickness $\Lambda = |\mathrm{Im}\{Q'_z\}|^{-1}$. For a finite t one has to worry, whether the observed Bragg scattering is dominated by the surface or the bulk behaviour of the OP: When we remember that surface modified critical behaviour is detectable within a surface layer of thickness ξ (4.1), we conclude that the crossover from bulk to surface critical scattering occurs when $\xi \simeq \Lambda$, thus, when

$$t \simeq t_{\mathrm{sb}} = \left(\frac{\xi_0^-}{\Lambda}\right)^{1/\nu} \quad . \tag{4.14}$$

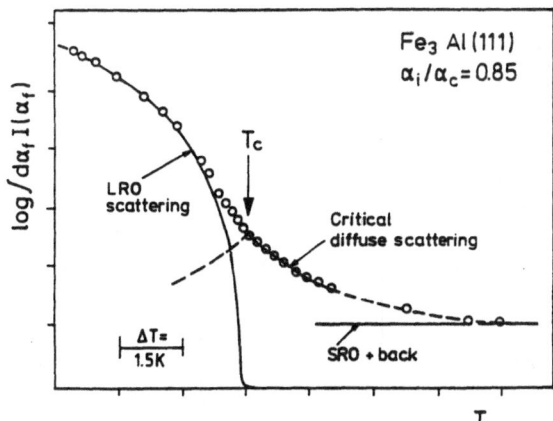

Fig. 4.5. General temperature dependence of the (111) evanescent x-ray intensity at the $Fe_3Al(1\bar{1}0)$ surface. The open symbols are experimental values [Mailänder et al., 1991]; the full line indicates schematically the near-surface LRO, the dashed curve the near-surface critical diffuse scattering

In other words, the more surface sensitive one renders the x-ray or neutron probe, the more extended in temperature is the surface-critical regime. This is one of the major differences to the SPLEED measurements discussed above which "see" only one critical regime. Here one has within the conventional critical regime a surface-critical regime whose extension depends on the scattering conditions.

This x-ray scattering experiment has been performed at the $Fe_3Al(1\bar{1}0)$ surface which contains the (111) superlattice reflection [Mailänder et al., 1991]. The experimental setup (W1-beamline at HASYLAB/Hamburg) was similar to that discussed in Sect. 2.4.3 (Fig. 2.14). For a fixed incidence angle $\alpha_i/\alpha_c \simeq 0.9$ the temperature dependence of the α_f-profiles of the evanescent superlattice peak was determined. The general temperature dependence of the scattered intensity evaluated around $\alpha_f/\alpha_c = 1$ is summarized in Fig. 4.5. The full line indicates how the surface LRO disappears with $t \to 0_-$, the dashed curve shows the schematic development of the diffuse scattering arising from critical near-surface OP fluctuations (which will be discussed below). The critical diffuse scattering around the superlattice vector is strongest at T_c. This criterion has been used for the experimental determination of T_c with an error of ± 0.5 K. Then, after the appropriate subtraction of the critical diffuse scattering below T_c, the powerlaw behaviour of the LRO scattering can be analyzed and gives (Fig. 4.6a)

$$\beta_1(Fe_3Al) = 0.75 \pm 0.02 \tag{4.15}$$

which again confirms the theory as far as the values of β_1 is concerned.

I want to add at this point some comments on the size of the critical region for β_1. It has to be deduced from the deviation from the asymptotic behaviour of $m(z, t)$ (4.6). Within RG theory to $O(\varepsilon)$ one gets [Gompper, 1984]:

$$f^*(x, 0, 0) = r_0(\varepsilon)x^{(\beta_1 - \beta)/\nu} - r_1(\varepsilon)x^{(\beta_1 - \beta + 1)/\nu} \tag{4.16}$$

with $r_0(\varepsilon), r_1(\varepsilon)$ universal. The associated Bragg intensity is again straightforward to calculate (using (B.3–5)):

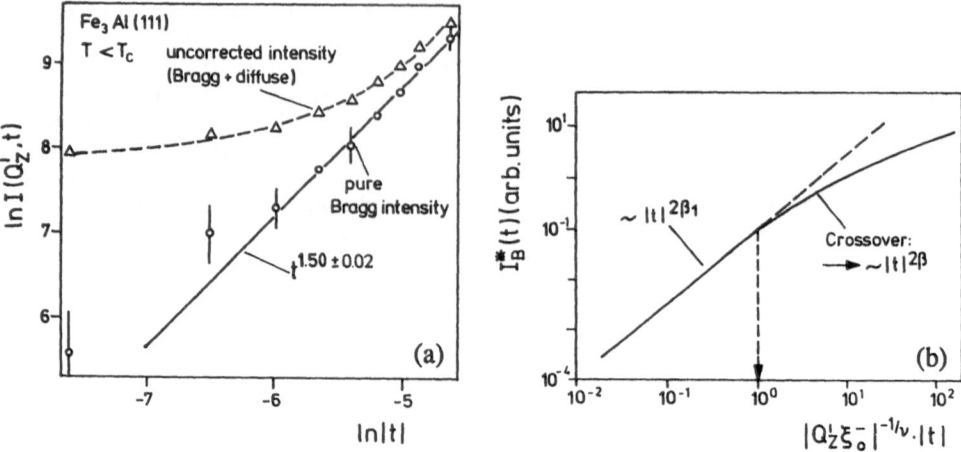

Fig. 4.6a,b. Temperature dependence of the near-surface LRO close to the bulk critical point. (a) x-ray experiment at the Fe$_3$Al(1$\bar{1}$0) surface [Mailänder et al., 1991]; (b) renormalization group calculation [Gompper, 1984]

$$I_B^*(Q_z', t \to 0_-) = I_B(Q_z', t \to 0_-)\left|1 - R_B(-iQ_z'\xi)\right|^2 \quad , \tag{4.17}$$

where $I_B(Q_z', t)$ is given by (4.13) and

$$R_B(y) \propto y^{-1/\nu} \tag{4.18}$$

is a correction term to the asymptotic form (4.13). With $|1 - R_B|^2 \simeq 1 - 2\,\mathrm{Re}\{R_B\}$ one concludes that the leading deviation from the asymptotic behaviour is governed by $\mathrm{Re}\{R_B\} = |Q_z'\xi|^{-1/\nu}\cos[1/\nu(\arccos\zeta)]$ which drops out for $\zeta = \zeta_B^* = \cos(\nu\pi/2) \simeq 0.5$ (using $\nu = 0.67$). The *"degree of evanescence"* ζ,

$$\zeta \equiv \frac{\mathrm{Im}\{Q_z'\}}{|Q_z'|} \quad , \tag{4.19}$$

is 1 for ideal evanescent waves (i.e. $\alpha_i/\alpha_c \leq 1$, $\alpha_f/\alpha_c \leq 1$ and $\mu = 0$) and less than 1 otherwise. This quantity which appears here has been introduced by Dietrich and Wagner [1984]. It plays an important role in the theory of critical surface scattering and is of considerable practical importance, as here, where the proper choice of ζ alloys the experimental minimization of intensity correction terms. The grazing angles $\alpha_i/\alpha_c = 0.9$ and $\alpha_f/\alpha_c \simeq 1$ which have been used in the experiment by Mailänder et al. [1991] approximately meet the optimal condition $\zeta = \zeta_B^*$ where the surface-critical behaviour is largest in t. For this degree of evanescence Gompper [1984] has calculated the variation of the Bragg scattering with the quantity $|\mu|^{-1/\nu} = |iQ_z'\xi_0^-|^{-1/\nu}|t|$ (Fig. 4.6b) and predicts a crossover at $|\mu|^{-1/\nu} \simeq 1$ from the pure $|t|^{-2\beta_1}$ powerlaw to a more complicated behaviour which is essentially governed by the bulk critically ($\simeq |t|^{2\beta}$). One can easily convince oneself the $|\mu|^{-1/\nu} \simeq 1$ corresponds essentially to $t = t_{sb}$ (4.14) which we derived from very intuitive arguments. Inserting $\nu = 0.67$, $\xi_0^- = 0.6\,\text{Å}$

[Guttman and Schnyders, 1969] and $\Lambda = 120$Å, one concludes that the evanescent Bragg scattering should only be sensitive to β_1 for $t \leq t_{sb} = 3 \times 10^{-3}$. Although the experimental t-regime $9 \times 10^{-4} \leq |t| \leq 6 \times 10^{-3}$ extends only marginally to t_{sb} without exhibiting deviations from the asymptotic β_1-powerlaw, this could indicate that the surface-critical regime is, for some unknown reason (e.g. surface-enhanced amplitude ξ_0^-), more extended than estimated. The same trend may also be noted in the experiments at the Ni(100) surface, where the critical region appears remarkably large (Fig. 4.2).

4.2 Surface Induced Decay of Critical Correlations

While the average value of the OP vanishes continuously upon approaching the critical temperature, its local fluctuations from its average value grow stronger, until at T_c, OP fluctuations occur at any length scale. In the bulk of a critical system this phenomenon is described by a pair correlation function of the form

$$g(\boldsymbol{r}_i - \boldsymbol{r}_j) \equiv \langle (m_i - \langle m \rangle)(m_j - \langle m \rangle) \rangle \propto \frac{e^{-|r_i - r_j|/\xi}}{|\boldsymbol{r}_i - \boldsymbol{r}_j|^{1+\eta}} \tag{4.20}$$

which disappears exponentially fast far away from T_c. Close to T_c the pair correlation function decays with a powerlaw as given by the exponent $1 + \eta$ ($\eta \simeq 0.03 - 0.08$ is the Fisher exponent [Fisher and Burford, 1967]) and ranges all length scales provided by the system. The form of $g(r)$ implies a critical diffuse scattering intensity (with $q \equiv Q - G_{hkl}$)

$$I_{\text{dif}}(q) \propto (q^2 + \xi^{-2})^{(\eta/2)-1} \tag{4.21}$$

which exhibits a pronounced divergence at $t = 0$ as $q \to 0$ ("*critical opalescence*"). In a pioneering neutron scattering study Als–Nielsen and Dietrich [1967] analyzed the shape of this critical scattering in β-CuZn and concluded that the numerical value of η could only be pinned down to the range $\eta \simeq [-0.05, +0.10]$. A more precise assessment of η was possible with the use of the scaling relation $\gamma = \nu(2 - \eta)$ yielding $\eta = 0.077 \pm 0.067$. In an x-ray study of Fe$_3$Al Guttman et al. [1969] observed critical diffuse scattering which is consistent with $\eta = 0.080 \pm 0.005$. One of today's most accurate experimental values of the elusive exponent η results from a neutron small angle scattering study of a critical binary liquid which gives $\eta = 0.0858 \pm 0.0263$ [Schneider et al., 1980].

The presence of the surface leads to a severe disturbance in the decay of the OP correlation function: Due to the loss of translational invariance at the two-point correlations do no longer depend on the distance between two points $\boldsymbol{r}_i$ and $\boldsymbol{r}_j$, but also on their distances z_i and z_j from the surface. A closed form of $g(\boldsymbol{r}_i, \boldsymbol{r}_j)$ for arbitrary t is not known, however, at $t = 0$ one finds

$$g(\boldsymbol{r}_i, \boldsymbol{r}_j, z_i, z_j) \propto \begin{cases} |\boldsymbol{r}_i - \boldsymbol{r}_j|^{-(1+\eta_\parallel)} & \text{for } |\boldsymbol{r}_i - \boldsymbol{r}_j| \to \infty; \quad z_i, z_j \text{ fixed} \\ |\boldsymbol{r}_i - \boldsymbol{r}_j|^{-(1+\eta_\perp)} & \text{for} \qquad\qquad z_j \to \infty; \quad z_i \text{ fixed} \\ |\boldsymbol{r}_i - \boldsymbol{r}_j|^{-(1+\eta)} & \text{for } |\boldsymbol{r}_i - \boldsymbol{r}_j| \to \infty; \quad z_i, z_j \to \infty \end{cases}$$

$$(4.22)$$

with the *universal* surface-induced critical exponents [Reeve and Guttman, 1980; Diehl and Dietrich, 1981]

$$\eta_\parallel \simeq 1.48 \,, \quad \eta_\perp \simeq 0.76 \tag{4.23}$$

which describe the asymptotic decay of the two-point correlations parallel and normal to the surface, respectively, which is, according to the numerical numbers (4.23), quite anisotropic. Note that $\eta_\parallel$ and $\eta_\perp$ are related by the scaling relation $2\eta_\perp = \eta_\parallel + \eta$ [Diehl and Dietrich, 1981]. Since $\eta \simeq 0$, this leads to the rule of thumb $\eta_\parallel \simeq 2\eta_\perp$. Two features of $g(\boldsymbol{r}_i, \boldsymbol{r}_j)$ appear quite remarkable:

a) The asymptotic decay of $g(\boldsymbol{r}_i, \boldsymbol{r}_j)$ parallel to the surface is governed by the surface exponent $\eta_\parallel$ no matter how far from the surface they occur (note that z_i, z_j are fixed, but otherwise arbitrary), in other words, at T_c the entire critical system behaves surface-like.

b) The large value of $\eta_\parallel$ expresses a dramatic action of the surface which induces a rapid spatial decay of the correlations of the near-surface OP fluctuations.

It was stressed by Dietrich and Wagner [1984] that the latter point gives birth to qualitatively new scattering phenomena when evanescent x-ray scattering is employed:

While bulk critical phenomena create divergences in the diffuse scattering, the surface-modified OP fluctuations at an ordinary transition only induce cusp-like scattering singularities near $t = 0$ and $q_\parallel = 0$. The evanescent critical diffuse scattering associated with $g(\boldsymbol{r}_i, \boldsymbol{r}_j)$ is

$$I_{\text{dif}}(\boldsymbol{q}_\parallel, Q_z') \propto \int_0^\infty \int_0^\infty g(\boldsymbol{q}_\parallel, z_i, z_j) \, e^{i(Q_z' z_i - Q_z'^* z_j)} dz_i \, dz_j \tag{4.24}$$

$$g(\boldsymbol{q}_\parallel, z_i, z_j) = \int g(r_{ij}, z_i, z_j) e^{i\boldsymbol{q}_\parallel \cdot \boldsymbol{r}_{ij}} d^2\boldsymbol{r}_{ij} \quad, \tag{4.25}$$

where $\boldsymbol{r}_{ij} \equiv (\boldsymbol{r}_i - \boldsymbol{r}_j)_\parallel$ and $\boldsymbol{q}_\parallel \equiv \boldsymbol{Q}_\parallel - \boldsymbol{G}_{hkl}$ ($\boldsymbol{G}_{hkl}$ lies in the surface). The asymptotic limits ($t = 0$; $q_\parallel \to 0$) and ($q_\parallel = 0$; $t \to 0$) of $I_{\text{dif}}(\boldsymbol{q}_\parallel, Q_z')$ can be deduced from scaling expressions for $g(\boldsymbol{q}_\parallel, z_i, z_j)$ to be (see also Gompper, [1986])

$$I_{\text{dif}}(\boldsymbol{q}_\parallel \to 0, Q_z')\Big|_{t=0} \simeq |-iQ_z'|^{\eta-3} A_0(\zeta) \tag{4.26}$$

$$\times \left(1 + B_1(\zeta)\Xi^{\eta_\parallel - 1} + B_2(\zeta)\Xi^2 + B_3(\zeta)\Xi^{\eta_\parallel+1} + \ldots\right)$$

$$I_{\text{dif}}(Q_z', t \to 0_\pm)\Big|_{q_\parallel=0} \simeq |-iQ_z'|^{\eta-3} A_0(\zeta) \tag{4.27}$$

$$\times \left(1 + A_1^\pm(\zeta)\Theta^{-\gamma_{11}} + A_2^\pm(\zeta)\Theta + A_3(\zeta)\Theta^{1-\gamma_{11}} + \ldots\right)$$

with

$$\Xi \equiv \left| \frac{q_\parallel}{iQ'_z} \right| \quad , \qquad \Theta \equiv \left| iQ'_z \xi_0^\pm \right|^{-1/\nu} |t| \tag{4.28}$$

and ζ (4.19) as before. The amplitudes $A_{0123}(\zeta)$ and $B_{123}(\zeta)$ have been determined for the ordinary fixpoint within MF theory [Dietrich and Wagner, 1984] and within RG theory [Gompper, 1986]. Since they depend on the degree of evanescence ζ, they can be taylored in a certain range by the proper choice of the grazing angles, a degree of freedom which can be used in a grazing angle scattering experiment to force the coefficient $A_2(\zeta)$ and $B_2(\zeta)$ of the leading correction term to zero. For this mere technical purpose the MF expressions,

$$A_2(\zeta) = B_2(\zeta) = 4\zeta^2 - 1 \tag{4.29}$$

are accurate enough which drop out of (4.26,27) for $\zeta^* = 0.5$. By noting that $A_1(\zeta) < 0$ and $B_1(\zeta) < 0$ one follows that the leading terms in (4.26,27) assume a cusp-like shape as a function of $q_\parallel$ and t, respectively (Fig. 4.7). Apparently, the surface-induced decay of the critical two-point correlations is so efficient that the critical divergence of the diffuse scattering is destroyed.

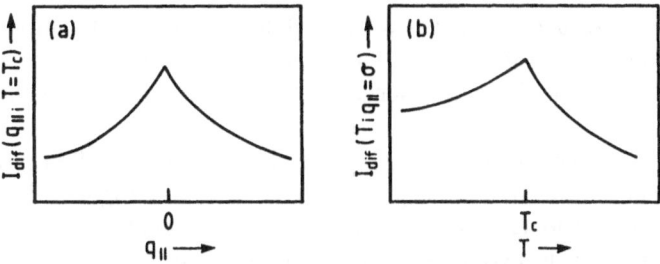

Fig. 4.7a,b. Cusp-singularities of near-surface critical diffuse scattering (schematically); (a) diffuse evanescent intensity versus $q_\parallel$ at $t = 0$ (*symmetric cusp*); (b) diffuse evanescent intensity versus t at $q_\parallel = 0$ (*asymmetric cusp*)

This very distinct theoretical prediction has been subject to an experimental test by Mailänder et al. [1990a,1990b] at the ($1\bar{1}0$) surface of the system Fe_3Al. It turned out that the unambiguous experimental verification of the existence of a cusp-like diffuse scattering poses some serious experimental challenges: Since the cusp-like feature is only present for $t = 0$, an accurate experimental determination of T_c and a precise temperature control and long time temperature stability of the sample surface is necessary. The occurence of a diffuse cusp instead of a diffuse divergence (as in the bulk) means, as well shall see, a small scattering cross section which has to be observed under grazing angle scattering conditions. In addition there are narrow constraints in the choice of the evanescent waves (via the quantity ζ) which in turn require a precise control of the grazing incidence and exit angles. The actual experiment consists of a repeated line-shape analysis

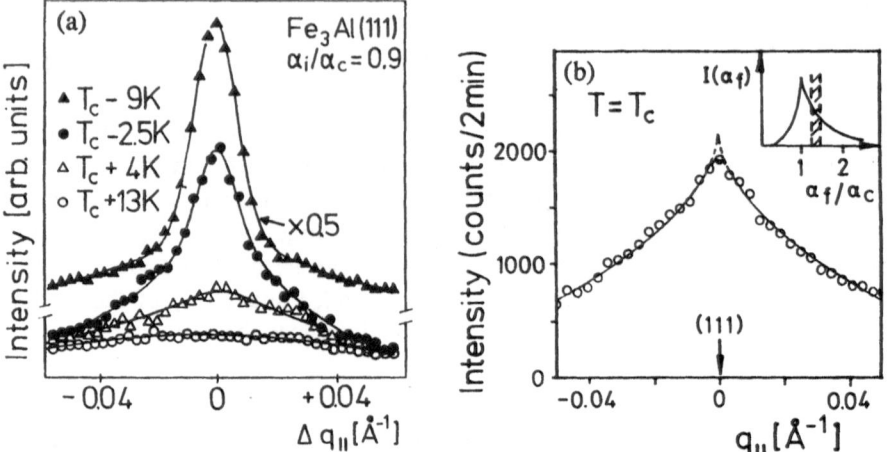

Fig. 4.8a,b. Evanescent diffuse scattering around the (111) reciprocal lattice vector at the $Fe_3AL(1\bar{1}0)$ surface observed at $\alpha_i/\alpha_c = 0.9$ and $\alpha_f/\alpha_c = 1.3$ [Mailänder et al., 1990]; **(a)** temperature dependence of the intensity distribution; **(b)** observation at $t = 0$ ($\pm 1 \cdot 10^{-3}$) together with a theoretical fit including instrumental resolution (full line). The dashed curve shows the deconvoluted cusp-scattering law

of the diffuse scattering around the (111) reciprocal lattice vector at various temperatures around the bulk critical point, thus, one needs a well-defined x-ray beam and a mirror-like single-crystal surface which do not smear out the expected profile.

The experiment by Mailänder et al.[1990a] has been carried out at HASYLAB using the brilliant synchrotron x-radiation provided by a 32-pole wiggler. The incident energy was choosen to $E_i = 6.90$ keV below the K-absorption edge of Fe ($E_K = 7.13$ keV) in order to avoid the strong Fe fluorescence scattering. α_i was kept fixed throughout the experiment at $0.9\,\alpha_c$ ($\alpha_c = 7.2$ mrad), while the diffuse scattering was recorded in an α_f-intervall between 0 and $4\,\alpha_c$ by the use of a PSD. The scattering condition $\zeta = \zeta^* = 0.5$ requires that $\kappa'_z = \text{Re}\{Q'_z\} \neq 0$, thus $\alpha_f > \alpha_c$. With $\alpha_i = 0.9\,\alpha_c$ it follows that $\alpha_f/\alpha_c \simeq 1.25\,\alpha_c$ meets the optimal condition for the observation of the diffuse cusp which is well within the α_f-range provided by the PSD. Some of the measured diffuse intensities profiles $I_{\text{dif}}(q_{\parallel}\,;\,\alpha_i = 0.9\,\alpha_c\,,\,\alpha_f = 1.3\,\alpha_c)$ around T_c are shown in Fig. 4.8a which shows Bragg intensity and strong diffuse intensity below T_c and softly decaying diffuse intensity above T_c. The evanescent critical diffuse scattering at $t = 0$ (Fig. 4.8b) displays indeed a very pronounced cusp-like line profile which can be used to obtain an experimental value of $\eta_{\parallel}$. By a least squares-fit analysis of the data with a profile function of the form

$$I(q_{\parallel}^*) \propto \int_{-\infty}^{\infty} \left(I_{\text{BG}} + \left(1 - I_0\,|q_{\parallel}|^{\eta_{\parallel}-1}\right)\right) e^{-(q_{\parallel}-q_{\parallel}^*)^2/\Delta q_{\parallel}^2}\,dq_{\parallel} \qquad (4.30)$$

the authors obtained a best fit (full line in Fig. 4.8b) for

$$\eta_{\parallel} = 1.52 \pm 0.04 \tag{4.31}$$

In (4.30) I_{BG} (constant scattering background), I_0 and $\eta_{\parallel}$ are free fitting parameters. The radial resolution $\Delta q_{\parallel}$ was essentially determined by the detector slit which was relaxed to $\Delta(2\theta) = 1.25$ mrad (σ-value) for intensity reasons, thus, $\Delta q_{\parallel} \simeq Q_{111} \cot(2\theta) \Delta\theta \simeq 2.7 \times 10^{-3} \, \text{Å}^{-1}$ was used in (4.30). This grazing angle x-ray scattering experiment confirms the exponent $\eta_{\parallel}$ as deduced from RG theory as well as the predicted lineshape of the evanescent critical diffuse scattering. Since the experimental value of $\eta_{\parallel}$ is very close to 3/2 and $\nu = 2/3$, one could naively assume the relation $\eta_{\parallel} = \nu^{-1}$ which has indeed been proposed by Bray and Moore [1977], incorrectly, as it turned out later ([Diehl and Dietrich, 1981]; see Sect. 4.3).

It was estimated by the author that the primary intensity intercepted by the Fe_3Al surface (1 cm^2) at $\alpha_i/\alpha_c = 0.9$ was approximately $10^9 - 10^{10}$ photons/s, while the observed cusp intensity was as weak as 10 counts/s (Fig. 4.8b). The conclusion from this is that such experiments appear only feasible with the use of highly brilliant wiggler or undulator radiation (as provided in modern synchrotron radiation laboratories). Albeit it would be desirable to get further experimental evidence for $\eta_{\parallel}$ from different systems belonging to other universality classes, in particular from magnetic surfaces, the chances to be able to observe a "magnetic" diffuse cusp with evanescent neutron scattering are very low.

The temperature dependence of the critical diffuse scattering at $q_{\parallel} = 0$ has the form (4.27) which is governed by the local susceptibility $\chi_{11} \equiv \delta m_1/\delta h_1 \propto |t|^{-\gamma_{11}}$ (at an order-disorder transition h_1 is a surface staggered field). In the case of an ordinary transition RG theory predicts $\gamma_{11} = -0.34$ (see Diehl [1983]), and accordingly χ_{11} remains finite at T_c (see Fig. 4.7b) in contrast to the diverging bulk quantity $\chi \equiv \delta m/\delta h = |t|^{-\gamma}$ ($\gamma = 1.25$). The form of $I_{dif}(q_{\parallel} = 0, Q'_z; t)$ (4.27) suggests a plot of the observed critical diffuse scattering for $T \geq T_c$ as

$$\ln\left(1 - \frac{I_{dif}(q_{\parallel} = 0, Q'_z, t)}{I_{dif}(q_{\parallel} = 0, Q'_z, 0)}\right) \propto -\gamma_{11} \ln|t| \tag{4.32}$$

which then should give the new surface exponent γ_{11}. Experimental results from the $Fe_3Al(1\bar{1}0)$ surface are available in a limited t-range $5.4 \times 10^{-3} \leq t \leq 2.5 \times 10^{-2}$ (Fig. 4.9), nonetheless the data allow a reliable determination of γ_{11}: The full line gives [Mailänder et al., 1991]

$$\gamma_{11} = -0.33 \mp 0.03 \tag{4.33}$$

which has to be compared with the theoretical value $\gamma_{11} = -0.34$ from ε-expansion.

How do the experimental $q_{\parallel}$ and t ranges, where surface critical diffuse scattering has been observed, compare with theoretical expectations for the associated critical ranges? We have to consider the correction terms to the leading behaviour (4.26,27) which are (see Gompper [1984,1986])

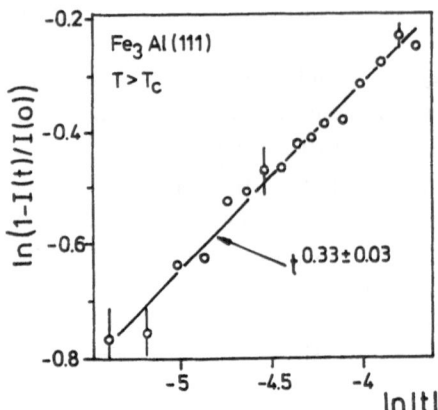

Fig. 4.9. Temperature dependence of the evanescent scattering at the (111) reciprocal lattice vector of the $Fe_3Al(1\bar{1}0)$ surface as observed for $t > 0$ [Mailänder et al., 1991]

$$I_{\text{dif}}\left(q_{\parallel} \to 0, Q'_z\right)\big|_{t=0} \simeq \left|-iQ'_z\right|^{\eta-3} A_0(\zeta^*) \tag{4.34}$$
$$\times \left(1 + B_1(\zeta^*)\varXi^{\eta_{\parallel}-1}[1 - R_{\parallel}(\zeta^*)\varXi^2 + \ldots]\right)$$

$$I_{\text{dif}}\left(Q'_z, t \to 0_{\pm}\right)\big|_{q_{\parallel}=0} \simeq \left|-iQ'_z\right|^{\eta-3} A_0(\zeta^*) \tag{4.35}$$
$$\times \left(1 + A_1^{\pm}(\zeta^*)\Theta^{-\gamma_{11}}[1 - R_t(\zeta^*)\Theta + \ldots]\right)$$

with $R_{\parallel}(\zeta^*) = -B_3(\zeta^*)/B_1(\zeta^*)$ and $R_t(\zeta^*) = -A_3(\zeta^*)/A_1(\zeta^*)$. Deviations from the leading powerlaws would accordingly not be expected, as long as $R_{\parallel}(\zeta^*)\varXi^2$ and $R_t(\zeta^*)\Theta$ are small compared to 1. Neglecting $\kappa'_z = \text{Re}\{Q'_z\}$ we get

$$|q_{\parallel}| < \frac{1}{\Lambda(R_{\parallel}(\zeta^*))^{1/2}} \qquad \text{for the } q_{\parallel}\text{-cusp} \quad , \tag{4.36}$$

$$|t| < \frac{|\xi_0/\Lambda|^{1/\nu}}{R_t(\zeta^*)} \qquad \text{for the } t\text{-cusp} \quad . \tag{4.37}$$

Since within MF theory $R_{\parallel}(\zeta^*) = R_t(\zeta^*) = 1$ (see Dietrich and Wagner [1984]), the MF version of the surface-critical regimes becomes rather simple, in particular we recover again our familiar result (4.14). For details of a more rigorous estimate within RG theory I refer the reader to the work of Gompper [1984,1986]: Including $O(\varepsilon)$ one then finds that $R_{\parallel}(\zeta_{\parallel}^*) = 0.52$ and $R_t(\zeta_t^*) = 0.43$ meaning that the critical $|t|$-regime now becomes twice as large as the associated MF-regime, while the critical $|q_{\parallel}|$-regime is not much affected[2]. By feeding the experimental parameters as reported by Mailänder et al.[1990a] into (4.36) we estimate that the $q_{\parallel}$-cusp should be visible within $|q_{\parallel}| \leq 2 \times 10^{-2}\,\text{Å}^{-1}$ which has to be compared with experimental regime $|q_{\parallel}| \leq 4 \times 10^{-2}\,\text{Å}^{-1}$ (Fig. 4.8a). The t-cusp is expected from (4.37) to emerge within $|t| \leq 1.3 \times 10^{-3}$, while experimentally it is observed for $T \geq T_c$ in the temperature range $t \simeq [5.4 \times 10^{-3}, 2.5 \times 10^{-2}]$. Both critical regimes for surface critical diffuse scattering are apparently somewhat larger than estimated.

---

[2] Including $O(\varepsilon)$ the amplitudes $A_2(\zeta)$ and $B_2(\zeta)$ of the leading correction terms vanish at $\zeta_t^* = 0.43$ and $\zeta_{\parallel}^* = 0.60$, respectively.

4.3 The Exponent ν and Surface Scaling Relations

So far we have discussed two asymptotic properties ($t = 0$, $q_{||} \to 0$ and $q_{||} = 0$, $t \to 0$) of the two-point correlation function, but we expect the two-point correlation function to behave smoothly away from "*asymptotia*": Figure 4.8a demonstrates how the $q_{||}$-profile gradually develops from a flat diffuse background away from $t = 0$ to the distinct cusp at $t = 0$. As in the bulk this metamorphosis of the diffuse scattering is driven by the diverging correlation length ξ (4.1). Hence the quantitative analysis of the temperature dependence of the diffuse wings should disclose the value of the exponent ν. Expanding the continuous behaviour of the diffuse scattering in an analytical power series one gets [Gompper, 1984]

$$I_{\text{dif}}(q_{||}, Q'_z, t)\Big|_{\substack{t \to 0 \\ q_{||} \to 0}} \simeq |-iQ'_z|^{\eta-3} A_0(\zeta) \tag{4.38}$$

$$\times \left(1 + A_1^{\pm}(\zeta)|\Theta|^{-\gamma_{11}}[1 + C_1(q_{||}\xi)^2 + C_2(q_{||}\xi)^4 + \ldots] + \ldots\right)$$

for the $q_{||}$-dependence of the diffuse scattering away from $t = 0$. Not too close to T_c we may neglect $q_{||}$-terms higher than $O(q_{||}^2)$ which leads to the conclusion that

$$\varphi(Q'_z, t) \equiv \frac{\delta I_{\text{dif}}}{\delta(q_{||}^2)} \simeq |-iQ'_z|^{\eta-3} A_0 A_1^{\pm} C_1 \xi^2 |\Theta|^{-\gamma_{11}} \propto |t|^{-\theta} \tag{4.39}$$

with the exponent

$$\theta = \gamma_{11} + 2\nu . \tag{4.40}$$

As expected the temperature dependence of $\varphi(Q'_z, t)$ allows an experimental assessment of the exponent ν near the surface. By fitting the wings of the diffuse scattering in Fig. 4.8a to a $q_{||}^2$-law the derivative $\varphi(Q'_z, t)$ is readily obtained and shown in Fig. 4.10 as a function of t. Away from $t = 0$ the powerlaw (full line) yields [Mailänder and Dosch, 1991]

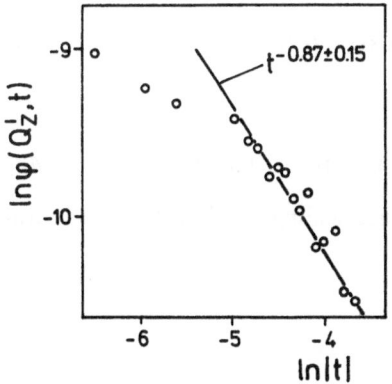

Fig. 4.10. The quadratic slope $\varphi(Q'_z, t)$ versus reduced temperature on a double-logarithmic scale [Mailänder and Dosch, 1991]

$$\theta = 0.87 \pm 0.15 \quad , \tag{4.41}$$

thus, inserting $\gamma_{11} = -0.33 \mp 0.03$, one finds

$$\nu = 0.60 \pm 0.15 \tag{4.42}$$

which is within the error bars indeed the bulk value. This is a very elegant experimental confirmation of a highly nontrivial fact:

The presence of a surface does not introduce a new length scale in near-surface critical phenomena ("Nothing matters except ξ").

In infinite critical systems the scaling properties of the free energy and the correlation functions imply the existence of 3 independent critical exponents, conventionally α, Δ and ν ("$\alpha - \Delta - \nu$–scaling"). It turns out that the ordinary transition in semi-infinite system requires only one additional surface exponent, say $\eta_{\parallel}$, all others follow from the scaling properties of the singular part of the surface free energy (see Diehl [1983])

$$f_s \propto |t|^{2-\alpha_s} g_t \left(h\, |t|^{-\Delta}, h_1\, |t|^{-\Delta_1} \right) \tag{4.43}$$

and of the two-point correlation function

$$G(r, z_1, z_2, t) = r^{-(d-2+\eta)} g_r(r/\xi, z_1/r, z_2/r) \tag{4.44}$$

(with $g_r(0, x, y) \sim x^{\eta_{\parallel} - \eta}$ for $x = y \to 0$). From (4.43) and (4.44) one follows e.g. that for $d = 3$

$$\beta_1 = \frac{\nu}{2}\left(1 + \eta_{\parallel}\right) \tag{4.45}$$

$$\gamma_{11} = \nu\left(1 - \eta_{\parallel}\right) \tag{4.46}$$

which are the counterparts of the bulk scaling relation $\beta = \nu(1 + \eta)/2$ and $\gamma = \nu(2 - \eta)$. With the 4 critical exponents β_1, γ_{11}, $\eta_{\parallel}$ and ν determined in the system Fe_3Al one is in the stage to subject these surface scaling relations associated with the ordinary transition to a first experimental test.

The experimental results deduced from critical evanescent x-ray scattering find for the relation (4.45)

$$0.75 \pm 0.02 = \frac{0.60 \pm 0.15}{2} [1 + (1.52 \pm 0.04)] = 0.76 \pm 0.15 \tag{4.47}$$

and for the relation (4.46)

$$-0.33 \mp 0.03 = (0.60 \pm 0.15)[1 - (1.52 \pm 0.04)] = -0.32 \pm 0.15 . \tag{4.48}$$

This experimental confirmation of the two surface scaling relations are a beautiful triumph for the semi-infinite RG theories.

4.4 Nonideal Stoichiometry, Surface Segregation and Dirty Surfaces

In the previous discussion of the critical scattering from Fe_3Al surfaces we pretended as if the sample and the sample surface were ideal. However, none of these tacit assumptions is fulfilled, in the contrary, one knows e.g.

a) that the composition of the Fe_3Al sample was actually $Fe_{0.71}Al_{0.29}$. (The sample was grown on purpose in this composition in order to get away from the miscibility gap around $c_{Al} = 0.25$, see Fig. 4.4a),
b) that extensive surface segregation of Al atoms occur at the $(1\bar{1}0)$ surface and
c) that the surface adsorbs gas atoms in the course of the experiments which lasted typically 2–3 weeks (!).

In order to make up leeway we now address these open questions, firstly, to which extent deviations from the ideal bulk stoichiometry and in particular surface segregation affect the interpretation of the observed critical scattering. What does one know from bulk systems? It is well established that the order-disorder transition in a noncongruent binary alloy (say AB) can be mapped onto an antiferromagnetic transition in a fixed external magnetic field g which couples to the average spin $\langle s \rangle$ (see e.g. [Bienenstock and Lewis, 1967]). The mapping yields that concentration difference $c_A - c_B$ corresponds to $\langle s \rangle$ and the difference in the chemical potentials $\Delta\mu \equiv \mu_A - \mu_B$ between the A and the B atoms adds to g. The renormalization of the bulk critical behaviour by such a noncritical ("hidden") variable which is subject to a constraint[3] has been discussed by Fisher [1968]. If g is not too strong, one finds that the critical temperature is lowered,

$$\Delta T_c \equiv T_c(0) - T_c(g) \propto \left(g\langle s \rangle\right)^2 \quad , \tag{4.49}$$

and the critical exponents get renormalized according to

$$\beta(g) = \frac{\beta(0)}{1 - \alpha'} \quad , \quad \gamma(g) = \frac{\gamma(0)}{1 - \alpha} \quad , \quad \alpha(g) = -\frac{\alpha(0)}{1 - \alpha} \quad . \tag{4.50}$$

(Note that $\alpha = 2\alpha' \simeq 0.12$). The general trend thus is that the diverging specific heat is turned finite by $g \neq 0$, while β and γ become slightly larger. The lowering of the critical temperature with increasing "field" g can be verified by inspection of the FeAl phase diagram, in fact, the renormalized bulk critical temperature of the Fe_3Al alloy used in the studies by Mailänder and coworkers has not been measured absolutely but determined to be $T_c \simeq 500$ K by the use of the phase diagram.

This *"hidden renormalization"* can readily be extended to semi-infinite systems [Wagner, 1984], here, however, an additional noncritical surface field g_1

---

[3] The introduction of a constraint is rather contrived but necessary to bring about the renormalization. In the case of a binary alloy, however, a natural constraint occurs, since $c_A + c_B = 1$.

("*surface magnetic field*") has to be considered which causes a surface enrichment of A or B atoms. It turns out that both fields g and g_1 have the same influence on the surface exponents as already known from bulk criticality: for $g \neq 0$ and/or $g_1 \neq 0$ one concludes that

$$\beta_1(x) = \frac{\beta_1(0)}{1 - \alpha'} \quad , \quad \gamma_{11}(x) = \frac{\gamma_{11}(0)}{1 - \alpha} \quad (x = g, g_1) \quad . \tag{4.51}$$

Since $\alpha' \simeq 0.06$, the renormalization of β_1 is virtually negligible, while $\gamma_{11}(0) \simeq -0.34$ gets renormalized to $\gamma_{11}(g) \simeq -0.36$ which still compares reasonably well with the experimental value $\gamma_{11} = -0.33 \mp 0.03$.

We consider the surface segregation mechanism in binary alloys in a little more detail (for a recent review see [Du Plessis, 1990]): A nonzero segregation field $g_1 \neq 0$ is in general caused by several mechanisms, as by the difference in surface tension of the alloy components and by the elastic stress energy owing to the difference in atomic size of the components. The resulting segregation profiles $c_A(z)$ can be very pronounced and depend on the competition between the surface field and the internal interactions (which is also surface dependent!). In particular for alloys which exhibit sublattice ordering (as Fe_3Al) $c_A(z)$ may strongly oscillate within the first layers (see e.g. [Gauthier and Baudoing, 1989]). The surface segregation kinetics in FeAl alloys have been studied by various authors (see [Gemmaz et al. [1990]). At the $(1\bar{1}0)$ surface a copious enrichment of Al takes place for temperatures $T \geq 400°$ C [Voges et al., 1992] which apparently leads to a corresponding Al-impoverishment in the second layer and so forth. This has some fundamental consequences for the near-surface critical behaviour and its observability with various surface techniques:

a) Since the almost complete supersession of the toplayer Fe atoms by Al totally destroys the order-disorder transition in the toplayer, the surface critical behaviour is hardly accessible to LEED. Here one of the trade-offs of evanescent x-ray (and neutron) scattering comes into play, because its sensitivity to subsurface layers can be tuned appropriately. The experiments by Mailänder et al. [1990a,b,1991] usually average over a surface layer of approximately 100 Å (see above).

b) What happens in the second layer, where an Fe enrichment is supposed to occur? The RG analysis of surface segregation starts assuming that the bulk is ideally stoichiometric, then any deviation from bulk stoichiometry in either direction leads to a lowering of the critical temperature. However, in the noncongruent $Fe_{0.71}Al_{0.29}$ alloy an Fe enrichment drives the system into the direction of the ideal Fe_3Al structure and should therefore increase LRO in this area, hence, also the associated critical temperature should increase. This surprising effect has first been observed by Mailänder et al. [1990a]: By tuning the evanescent x-rays most surface sensitive ($\alpha_i/\alpha_c = 0.9$ and $\alpha_f/\alpha_c = 0.8$ giving a scattering depth $\Lambda = 30$ Å), they could detect a tiny remaining LRO scattering for $T \geq T_c$ (bulk) (see Fig. 4.11) which the authors referred to this segregation phenomenon[4]. I

---

[4] Within the DWBA one can estimate that this Bragg scattering decays to about 1 count/3s when $\alpha_f/\alpha_c \geq 1.3$, thus, it is no longer visible in Fig. 4.8b underneath the critical diffuse cusp.

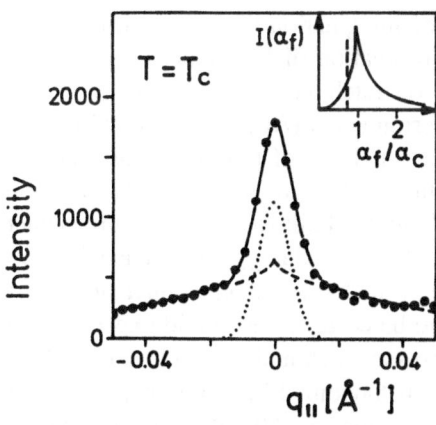

Fig. 4.11. Evanescent scattering intensity distribution around the (111) reciprocal lattice vector at the Fe$_3$Al(1$\bar{1}$0) surface observed at $t = 0$ ($\pm 1 \cdot 10^{-3}$) and $\alpha_i/\alpha_c = 0.9$ and $\alpha_f/\alpha_c = 0.8$ [Mailänder et al., 1990a]. The dotted curve shows the LRO contribution, the dashed curve the diffuse cusp from Fig. 4.8b

want to note already at this point that this surface segregation-mediated surface peak appears to occur also in other binary alloys as in Cu$_3$Au [Liang, 1991] (see Sect. 5.1.1).

From the theoretical viewpoint this situation is not well understood. There is one philosophy (mainly promoted by Bray and Moore [1977]) according to which it is rather unimportant, where a surface order above T_c (bulk) originates from, one should always observe extraordinary behaviour and the associated surface exponents in such a case, i.e. in particular $\eta_\parallel = 5$ (MF value). According to the observation in Fe$_3$Al this is apparently not the case. In a RG sense there exist two scaling fields which can induce surface order above T_c, the staggered surface field h_1 and the surface coupling parameter c (4.2), the latter being relevant only at the special transition, since small deviations from $c = c_{SP}$ decide whether an ordinary or an extraordinary transition occurs (Fig. 4.1b). For $c > c_{SP}$ surface order occurs in the presence of a disordered bulk which, however, necessarily induces an extraordinary transition. The scaling behaviour of a h_1 field is quite different, since according to (4.43) the efficiency of h_1 depends on the reduced temperature. The $h_1 - T$ (MF) phase diagram according to Nakanishi and Fisher [1983] is shown in Fig. 4.12 and suggests a crossover from ordinary to more complicated behaviour upon nearing T_c, whereby the crossover temperature moves to T_c as $h_1 \to 0$.

In the case of an order-disorder transition in a binary alloy (like Fe$_3$Al), however, the general wisdom has been that the associated h_1 field is always zero,

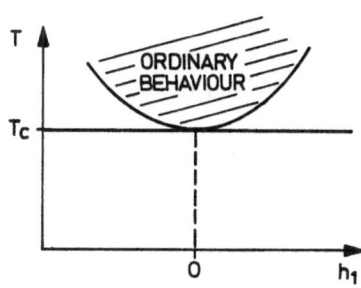

Fig. 4.12. $h_1 - T$ phase diagram (after Nakanishi and Fisher [1983])

since no demon was so far in the range of imagination who could bring about the correct staggered surface field. The observation of Mailänder et al. [1990a] indicates in which direction one may have to stretch one's imagination (translated into the *"Ising language"*): When an antiferromagnet (AF) in an external magnet field $g \neq 0$ is exposed to a magnetic surface field g_1 of opposite sign such that it cancels at the surface the g field, then the antiferromagnetic order at $z = 0$ is higher than in the bulk and may survive above the bulk Néel temperature (which is lowered according to (4.49) due to the hidden variable g). This means in other words, two antiparallel noncritical magnetic fields g and g_1 applied to an AF may produce an h_1 field. If this turns out to be correct, one would expect for the surface critical behaviour of Fe_3Al that a crossover should occur close to T_c from the observed ordinary to an extraordinary behaviour. Since ordinary behaviour was observed by Mailänder et al. [1990a] as close as ± 0.5 K to T_c, one would be forced in a future experiment to hunt within this small temperature regime for a possibly present crossover effect.

Which effect do gas adsorbates at the surface have on the surface-critical behaviour? Experimentally this is not investigated in a systematic way. It was reported by Mailänder and Dosch [1991] that during their two-weeks experiment they observed additional scattering apparently from gas atoms adsorbed at the Fe_3Al surface. They noticed, however, that the critical scattering did not change within the accuracy and sensitivity of their method. Diehl and Nuesser [1990] analyzed this situation theoretically by treating the "surface dirt" as a random surface field (*"quenched surface disorder"*, QSD). According to their conclusions the effect of dirty surfaces on near-surface critical behaviour is rather involved, i.e. it depends among else on the character of the bulk phase transition, on the surface universality class and on the degree of correlation within the QSD. However, quite generally it seems to hold that QSD is much less harmful than quenched bulk disorder simply for dimensional reasons. From this RG analysis it appears as if the exponent $\eta_\parallel$ (ordinary) which governs the spatial decay of the lateral critical fluctuations should be sensitive to QSD to some extent as soon as the QSD becomes long ranged correlated. When such a QSD is described by a random surface field of the form

$$g_{QSD}(r_\parallel) \propto r_\parallel^{-s} \tag{4.52}$$

then one finds from scaling arguments that the powerlaw decay (4.52) has to fulfill the condition

$$s > d - \eta_\parallel \quad , \tag{4.53}$$

if the associated surface dirt should be irrelevant. For $d = 3$ and $\eta_\parallel \simeq 1.5$ this means $s > 1.5$. In the experiment by Mailänder et al. [1990a] the adsorbed gas atoms apparently did not exhibit long ranging correlations with $s \leq 1.5$, however, it is not a priori evident what happens at a strongly oxidized sample surface.

4.5 Evanescent Magnetic Neutron Scattering from the MnF₂(001) Surface

So far the predictions of the RG theory of surface-critical behaviour have been tested at ferromagnets (Ni), binary alloys (Fe₃Al) and binary liquids [Sigl and Fenzl, 1986]. In order to test the universality hypothesis, experiments on the surface of antiferromagnets would be highly desirable. Interestingly, experiments on surface antiferromagnetism have already been performed in the late sixties on the NiO(100) surface using electron diffraction [Palmberg et al., 1968; Dewames and Wolfram, 1969], where it was convincingly demonstrated that the surface-antiferromagnetism decays faster than the bulk antiferromagnetic LRO. A re-analysis of the data by Al Usta and Dosch [1991] showed that they are reasonably compatible with a surface exponent $\beta_1 \simeq 0.8$, the effect of the critical diffuse scattering, however, has not been accounted for.

Another well known antiferromagnetic system is the insulator MnF₂: The magnetic moments are carried by the Mn⁺⁺ ions ("$S = 5/2$"-state; no orbital contribution) which are arranged on a body centered tetrahedral structure with $a = 4.87$ Å and $c = 3.31$ Å (see [Erickson, 1953]). In the ordered state ("*rutile structure*") the Mn spins are aligned parallel and antiparallel to the c-axis ($\equiv$ (001) direction) as shown in Fig. 4.13. In a neutron diffraction experiment one finds pure magnetic superlattice reflexions

$$F_{hkl}^m \propto 1 - e^{i\pi(h+k+l)} = 2 \quad \text{for } h + k + l = \text{odd integer}^5 . \tag{4.54}$$

The phase diagram of MnF₂ has been determined by specific heat [Teaney, 1965] and NMR measurements [Heller, 1966], by this the Néel temperature is located at $T_N = 67.5$ K. In the $T - H$ plane the phase diagram of MnF₂ consists of three phases, the paramagnetic (P), the antiferromagnetic (AF) and the spin-flop (SF) phase with a bicritical point at $T_b = 64.7$ K *and* $H_b = 118.5$ kOe (see e.g. [Shapira and Becerra, 1976]). The bulk critical phenomena have been extensively studied by many experimental techniques including inelastic neu-

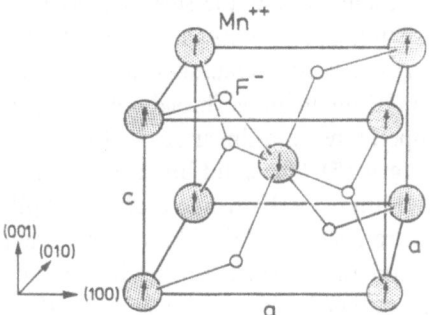

Fig. 4.13. Unit cell of MnF₂

---

⁵ Since $s_c \parallel$ (001), the (001) magnetic peaks have zero intensity.

tron scattering [Schulhof et al., 1970,1971] or magnetic birefringence[6] [Borovik–Romanov et al., 1971].

Let us consider the (001) surface. In the case of an ideal termination the Mn spins would be orientated perpendicular to the surface, since they are locked to the tetragonal (001) axis. Measurements of the thermal expansion by Gibbons [1959] show a strong temperature dependence of the c-axis below T_N and a distinct anomaly in the thermal expansion coefficient at T_N. It is a quite general effect that the cutoff of the surface bonds leads to a relaxation of the lattice parameter $a_\perp$ within the first few layers, therefore it is not unreasonable to assume that the spins near the (001) surface which are coupled to $a_\perp$ (= c-axis) will become partially disordered even well below the bulk Néel temperature. Thus, the experimental study of the surface-related antiferromagnetism in this system could disclose interesting thermodynamic phenomena.

Recently the antiferromagnetism at a $MnF_2(001)$ surface has been studied for the first time by neutron scattering under grazing incidence at the special instrument EVA of the Institut Laue–Langevin [Al Usta et al., 1992] which has been described in Sect. 3.4.2.

While the magnetic scattering is solely due to the magnetic moments of the Mn atoms, the occurence of total external neutron reflection is mediated solely by the F nuclei which have a positive coherent scattering length of $b_c^F = 5.65$ fm. Since the negative coherent scattering length $b_c^{Mn} = -3.73$ fm of the Mn nuclei tends to reduce the average scattering length density, the resulting neutron critical angle is only as small as $\alpha_c/\lambda = 0.784$ mrad/Å.

In the experiment the (100) antiferromagnetic surface Bragg reflection at the (001) MnF_2 surface has been studied in the temperature regime $T = 40$ K $- 67$ K. The incidence angle was choosen to be $\alpha_i/\alpha_c = 0.7$ in order to assure that the total incident neutron beam with a vertical divergence $\sigma_{iv}' = 1$ mrad impinges at the surface safely below the critical angle for total reflection which is $\alpha_c = 4.4$ mrad for the used neutron wavelength $\lambda = 5.5$ Å. A typical α_f-spectrum of the recorded Bragg scattering is shown in Fig. 4.14: The low absorption cross section allows the bulk Bragg scattering being transmitted through the entire crystal (see Sect. 3.4.2). As discussed in Sect. 3.4.2 the surface antiferromagnetic Bragg scattering can readily be identified. $\omega_\parallel$-scans through the surface peak (α_f-integrated) show the typical line-shape consisting of a two-component scattering law accounting for near-surface antiferromagnetic LRO (Gaussian line) and for diffuse evanescent neutron scattering due to OP fluctuations (broad component). The LRO contribution is plotted versus temperature on a linear scale (Fig. 4.15) and on a double logarithmic scale (inset in Fig. 4.15). The solid line is the best fit assuming a powerlaw behaviour, $m \propto t^\beta$, yielding $\beta \simeq 0.4$ which is close to bulk critical behaviour. The reason for this observation is quite clear: The α_f-integrated evanescent neutron scattering has only a moderate surface sensitivity, in this case $\Lambda \lesssim 1000$ Å. At the smallest achieved reduced temperature $t = 5 \times 10^{-3}$

---

[6] quadratic magneto-optic effect ($\equiv$ "quadratic Faraday–effect")

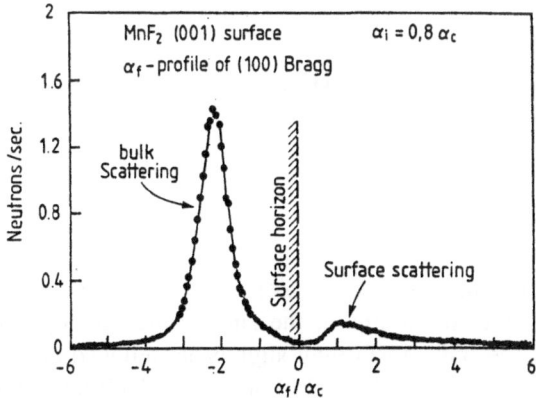

Fig. 4.14. α_f-spectrum of the (100) antiferromagnetic neutron Bragg scattering as observed at $\alpha_i/\alpha_c = 0.75$ [Al Usta et al., 1992]

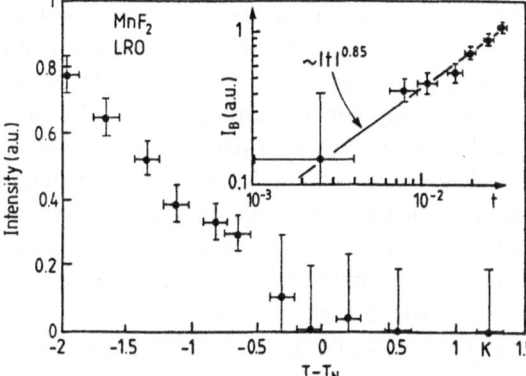

Fig. 4.15. Temperature dependence of the evanescent magnetic neutron scattering [Al Usta et al., 1992]: the inset shows the evanescent LRO contribution versus reduced temperature on a double logarithmic scale

the correlation length is $\xi \simeq 100$ Å, therefore $\xi/\Lambda < 1$, and one expects the evanescent critical neutron scattering to be dominated by a bulk critical exponent as observed. In order to become sensitive to surface critical behaviour, both the reduced temperature and the scattering depth has to be further reduced in future attempts.

5. Surface Effects
at First Order Phase Transitions

We consider now the surface of a solid which undergoes a discontinuous phase transition in the bulk. Assuming that the atoms at the surface are less tightly bound than their counterparts in the bulk, LRO may slightly be reduced at the surface. In this case the surface favours the disordered phase and facilitates the surface regime to reach the disordered state. As in the case of bulk-critical systems the characteristic feature of such surface-induced phenomena is a continuous spatial transition regime from the surface-modified behaviour to the pure bulk behaviour as one moves from the surface into the interior of the system, the length scale of this transition being determined by the correlation length ξ of the matter. In a bulk-critical system ξ diverges upon approaching criticality, therefore it is not too surprising that the action of the surface becomes particularly spectacular when the system approaches its critical point. Although the correlation length remains finite and of microscopic order in discontinuous phase transitions, the character of first order phase transitions, such as melting and most order-disorder transitions in binary alloys (for a review see Guttman [1956]), is fundamentally altered close to the border of the system, where surface-induced critical phenomena may occur which are completely absent in the bulk. The transition regime between the 2-dimensional surface and the 3-dimensional bulk then mediates between a *critical surface* and a *noncritical bulk*.

For a detailed discussion of the theory the reader is referred to the article by Lipowsky in a following volume of this series (see also the recent review by Dietrich [1988]).

5.1 Surface Induced Disorder in Cu₃Au

The bulk phenomena in Cu_3Au associated with the order-disorder transition at $T_0 = 663$ K (Fig. 5.1a) have been meticulously studied by the pioneers in the field: Warren [1969], Cowley [1950], Moss [1964,1966], Chipman [1956], and others. Thus, surface experiments on this alloy can be based on an exceptionally well understood bulk reference.

The ordered alloy exhibits a $L1_2$ structure with four simple-cubic sublattices, where three of them (α-sites) are occupied by Cu atoms and one (β-sites) by Au atoms (Fig. 5.1b). In the disordered state the lattice is randomly[1] occupied (fcc). The associated x-ray structure factors F_{hkl} are straightforward to calculate (see

---

[1] in a long range sense

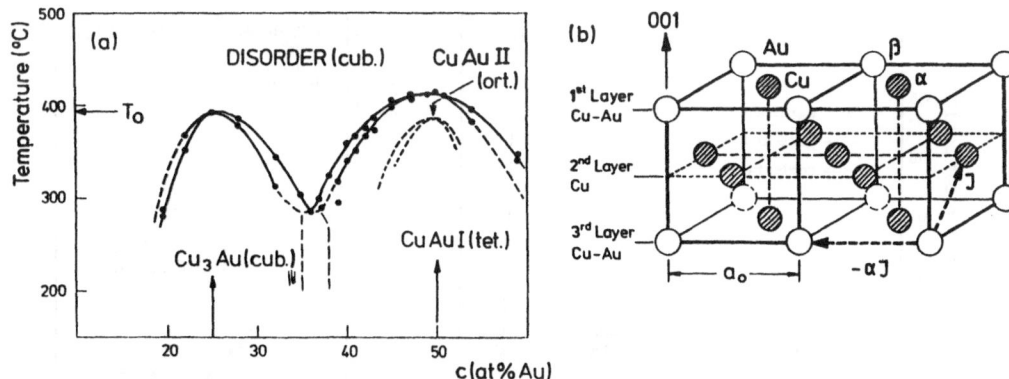

Fig. 5.1a,b. The binary alloy Cu-Au. (a) Part of the phase diagram (after Hansen and Anderko [1958]): (b) ordered Cu_3Au structure and associated (001) surface: $a_0 = 3.75$ Å, J = NN interaction, $-\alpha J$ = NNN interaction

Warren [1969]), in particular for (hkl) *mixed* one gets

$$F_{hkl} = \begin{cases} \propto (f_{Cu} - f_{Au}) & \text{for } L1_2 \\ 0 & \text{for } A2 \text{ (fcc)} \end{cases} , \qquad (5.1)$$

i.e. F_{hkl} (5.1) is a superlattice reflection which is sensitive to this order-disorder transition.

According to the experimental evidence the driving mechanism for this ordering can be understood within the Hume–Rothery rule: Measurements of the specific heat [Kuentzler, 1973] and a recent high resolution x-ray study of short range order (SRO) [Oshima et al., 1986] both suggest that the appearance of the new Brillouin zone in the ordered phase leads to a decrease in the electron density of states near the Fermi surface. However, it is also well known that a $L1_2$ structure cannot have a simple Ising ground state, if only "antiferromagnetic" interactions ($J > 0$) between the 12 nearest-neighbours (NN) are taken into account, thus, in order to remove the frustated "$T = 0$"-configuration, at least a competing "ferromagnetic" interaction among the 6 next-nearest-neighbours (NNN) is required, resulting in an Hamiltonian

$$H = J \sum_{NN} \tau_i \tau_j - \alpha J \sum_{NNN} \tau_i \tau_j , \qquad (5.2)$$

where $\tau_i = \pm 1$, when site i is occupied by Cu and Au, respectively, and $\alpha = 0.2$ [Binder et al., 1983]. A numerical estimate of the effective interaction energy J can be obtained within mean field theory from the transition temperature $T_0(MFA) = 4.18\, J/k_B$ (see e.g. [Gompper and Kroll, 1988] giving $J(MFA) = 13.7$ meV, i.e. $J = O(k_B T)$. Although the NNN interaction favours

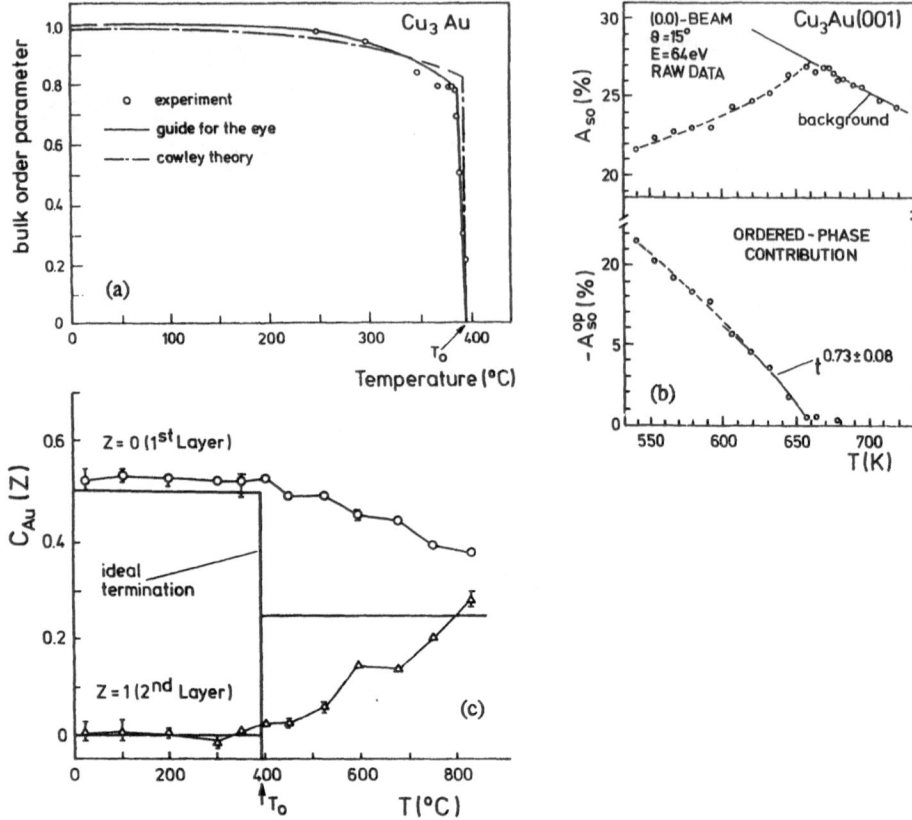

Fig. 5.2a-c. Order-disorder transition in Cu_3Au. (a) Temperature dependence of bulk LRO parameter [Cowley, 1950]; (b) temperature dependence of (001) surface LRO parameter [Alvarado et al., 1987]: raw data of SPLEED measurement (*top*), LRO contribution (*bottom*); (c) temperature dependence of Au concentration in the top layer and second layer of the (001) surface [Buck et al., 1983]

clustering, Cu_3Au does not exhibit any tendency towards phase separation, a fact which indicates that higher order and/or many body interactions may play a relevant role. As a consequence of this interlocked competing interaction, LRO in Cu_3Au does not disappear continuously upon approaching T_0, but instead, the order-disorder transition has been found by various x-ray scattering studies to be very pronounced first order (see Fig. 5.2a, Warren and Chipman [1949], Cowley [1950]).

5.1.1 Long Range Order (LRO) at the (001) Surface

At a free surface the atoms have a different, no longer close-packed, local environment, e.g. the atoms belonging to the top layer of the (001) surface (see Fig. 5.1b) have only 8 NN and 5 NNN (compared to 12 NN and 6 NNN in the bulk), and may therefore show a different ordering behaviour. Such a surface-induced effect on the LRO has indeed been discovered by LEED at the Cu_3Au

(001) surface [Sundaram et al., 1973,1974; McRae and Malic, 1984]: The top layer displays a very pronounced continuous order-disorder transition. Alvarado et al. [1987] showed with SPLEED measurements (Fig. 5.2b) that the surface OP m_1 obeys a power-law

$$m_1 = |t|^{\beta_1} \tag{5.3}$$

with $\beta_1 = 0.77$ ($t \equiv (T_0 - T)/T_0$ is the reduced temperature)[2]. Various theoretical approaches have emerged to explain this observation: Sanchez and Moran–Lopez [1985] analyzed an effective Hamiltonian of a semi-infinite fcc A_3B alloy within the cluster variation method and reproduced essentially the observed continuous behaviour of m_1 at the (001) surface, if the surface pair interaction $J_s = 0.72 J$ (J is the bulk value). One of the merits of this theoretical study is that the observed temperature dependence of both $m(z = 0, t)$ and $c_{Au}(z = 0, t)$ could be reproduced [Moran–Lopez et al., 1985]. Within this approach the continuous behaviour of m_1 appears as a 2-dimensional transition of a weakly coupled surface layer. A completely new insight into this kind of surface phenomena was provided by Lipowsky and co-workers [Lipowsky, 1982; Lipowsky and Speth, 1983; Lipowsky and Gompper, 1984; Lipowsky, 1987; Kroll and Lipowsky, 1983] who suggested that the continuous behaviour of m_1 is mediated by a wetting phenomenon driven by the discontinuous bulk phase transition: When the free surface of the ordered alloy favours disorder, an OP profile $m(z)$ as shown in Fig. 5.3a may develop at temperatures well below T_0. Within mean field theory it is straightforward to show that such a situation will finally lead to the emergence of a disordered surface layer of microscopic thickness $L(t)$ (Fig. 5.3b) which can grow to a mesoscopic and even macroscopic size when T_0 is approached. This implies a continuous temperature dependence of the surface OP m_1 with the *universal* quantity $\beta_1 = 0.5$. As long as short range interactions dominate the surface phenomena, one expects this wetting layer to grow according to

$$L(t) = \xi_d \ln |1/t| \quad , \tag{5.4}$$

where ξ_d is the correlation length in the disordered phase (which remains finite at all temperatures). However, this simple *universal* wetting scenario becomes considerably involved by the competition between the direct and the fluctuation-induced interactions and by the competition between ordering and non-ordering densitites. Both effects lead to the consequence that the surface phenomena associated with first order phase transitions in binary alloys (or in antiferromagnets) depend crucially on the system itself and on the surface under consideration. Although the OP fluctuations which have been ignored in the MFA play a very complicated role (see e.g. Binder et al. [1986]), one can say that they tend to increase β_1: Ignoring the effect of non-ordering densities one gets $0.5 \leq \beta_1 \leq 1$ [Lipowsky, 1987]. In binary alloys, however, the action of the aforementioned

---

[2] Note that the same notation β_1 is used which was originally introduced for semi-infinite critical systems.

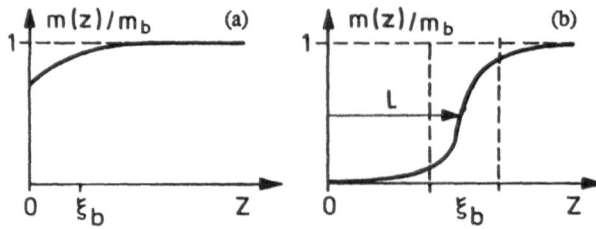

Fig. 5.3. OP profiles at a surface of a solid untergoing a 1st order transition (a) for $T < T_0$ and (b) for $T \leq T_0$. L denotes the average interface position

non-ordering densities comes into play that will be addressed now in some more detail: In analogy to the bulk Bragg–Williams order parameter (see e.g. Cowley [1950] we define the LRO near the surface as

$$m(z,T) = \frac{3}{4}\left(\frac{r_\alpha(z,T) - c_{Cu}(z,T)}{1 - y_\alpha(z)}\right) + \frac{1}{4}\left(\frac{r_\beta(z,T) - c_{Au}(z,T)}{1 - y_\beta(z)}\right) \qquad (5.5)$$

with c_{Cu} and c_{Au} as the actual Cu and Au concentrations ("*non-ordering densities*"), r_α and r_β as the fractions of the correctly occupied sites ("*ordering densities*") which range from 1 in the ordered state to c_{Cu} and c_{Au}, respectively, in the disordered state, and y_α and y_β as the fractions of the α- and β-sites. The non-ordering densities $c_{Cu}(z,T)$ and $c_{Au}(z,T)$ account for the actual composition of the near-surface layers which depend on the symmetry of the surface under consideration and may or may not be affected by surface segregation (in most cases they are). While in the bulk of the system c_{Au} and c_{Cu} are constant for a given system, they vary at the surface within the same length scale which governs the OP profile. This competition between ordering and non-ordering at the surface of A_3B systems renders the nature of the wetting transition rather complex, i.e. it depends particularly sensitively on the symmetry of the system and of the surface and on the range and type of the interactions.

The surface composition in the first and second layer of the Cu$_3$Au (001) has been measured by LEIS below and above the bulk transition temperature [Buck et al., 1983]: In the ordered state the equilibrium top layer is the Cu : Au $- 1 : 1$ layer followed by a pure Cu layer and so forth (Fig. 5.1b), thus, the (001) surface exhibits a segregation-free, ideally stoichiometric termination. The remarkable observation is that this layering remains unchanged at temperatures far beyond T_0 (Fig. 5.2c). Apparently, the order-disorder transition is not associated with an interdiffusion of the atoms between near surface layers. Theoretical values for the surface exponent β_1 deduced from a two-density mean field theory and from Monte–Carlo calculations [Gompper and Kroll, 1988; Kroll and Gompper, 1987] accounting for the layer composition range from 2.22 to 1.72 depending on the actual role of the interfacial fluctuations which is not yet well understood [Binder et al., 1986]. This range of β_1, however, contradicts the experimental value $\beta_1 = 0.77$ [Alvarado et al., 1987] which, to add to the confusion, seems to agree more with the predictions of a MF theory which discards the action of non-ordering densities (see e.g. Lipowsky [1987]).

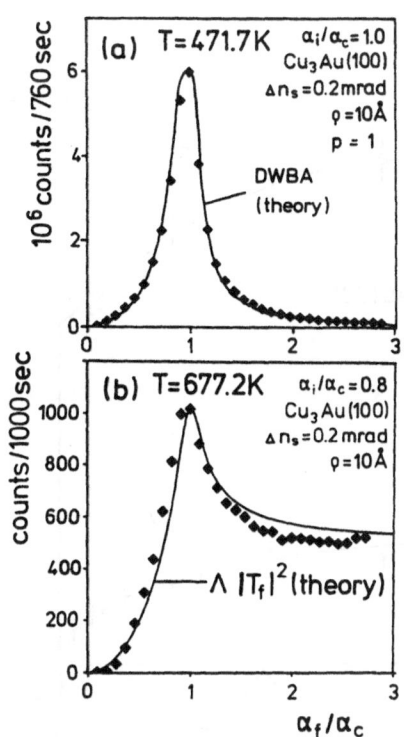

Fig. 5.4. α_f-profiles of evanescent (100) x-ray scattering at the Cu_3Au (001) surface for (a) $T = T_0 - 191.3K$ (Bragg scattering) and for (b) $T = T_0 + 14.2\,K$ (diffuse scattering). The solid lines are theoretical curves according to the DWBA with the surface parameter Δn_s, ϱ and p

Figure (a): $T = 471.7\,K$, $\alpha_i/\alpha_c = 1.0$, Cu_3Au (100), $\Delta n_s = 0.2\,mrad$, $\varrho = 10\,\text{Å}$, $p = 1$, DWBA (theory). Vertical axis: 10^6 counts/760 sec. Horizontal axis: α_f/α_c.

Figure (b): $T = 677.2\,K$, $\alpha_i/\alpha_c = 0.8$, Cu_3Au (100), $\Delta n_s = 0.2\,mrad$, $\varrho = 10\,\text{Å}$, $\Lambda\,|T_f|^2$ (theory). Vertical axis: counts/1000 sec. Horizontal axis: α_f/α_c.

While (SP)LEED is essentially sensitive to m_1, details of the wetting transition have been obtained by evanescent x-ray scattering [Dosch et al., 1988; Dosch and Peisl, 1989; Dosch et al, 1991a]: The experiment has been performed at the (001) surface of a Cu_3Au single crystal which was grown in a special Czochralsky apparatus [Uelhoff, 1987]. The evanescent scattering associated with the (100) reciprocal superlattice reflection has been investigated in the temperature range between room temperature (300 K) and 677 K (well above the transition temperature $T_0 = 663$ K). The authors recorded α_f-resolved spectra and α_f-integrated ω-scans of the (100) superlattice reflection at various incidence angles α_i around α_c. Complementary, the (200) scattering intensity was used as a watchdog for any unusual surface-related Debye–Waller factor and surface roughening effects associated with the order-disorder phase transition which were, however, not detected. The α_f-profile of the (100) superlattice peak (Fig. 5.4a) shows the characteristic asymmetric line shape typical for evanescent scattering [Dosch, 1987] which can fully be understood within the DWBA. The full line in Fig. 5.4a is calculated according to (2.40–45) taking $\varrho = 10\pm5\text{Å}$, $\Delta n_s = 0.2\,mrad$ and $p \simeq 1$. Figure 5.4b shows the α_f-profile of the diffuse scattering at the (100) reciprocal lattice point at $T = 677$ K $= T_0 + 14$ K and $\alpha_i/\alpha_c = 0.8$ and the theoretical curve (full line) using (2.38) and the surface parameters determined before: The pronounced peak at $\alpha_f/\alpha_c \simeq 1.0$ is solely due to the Vineyard enhancement of the interface and must not be confused with any remaining LRO.

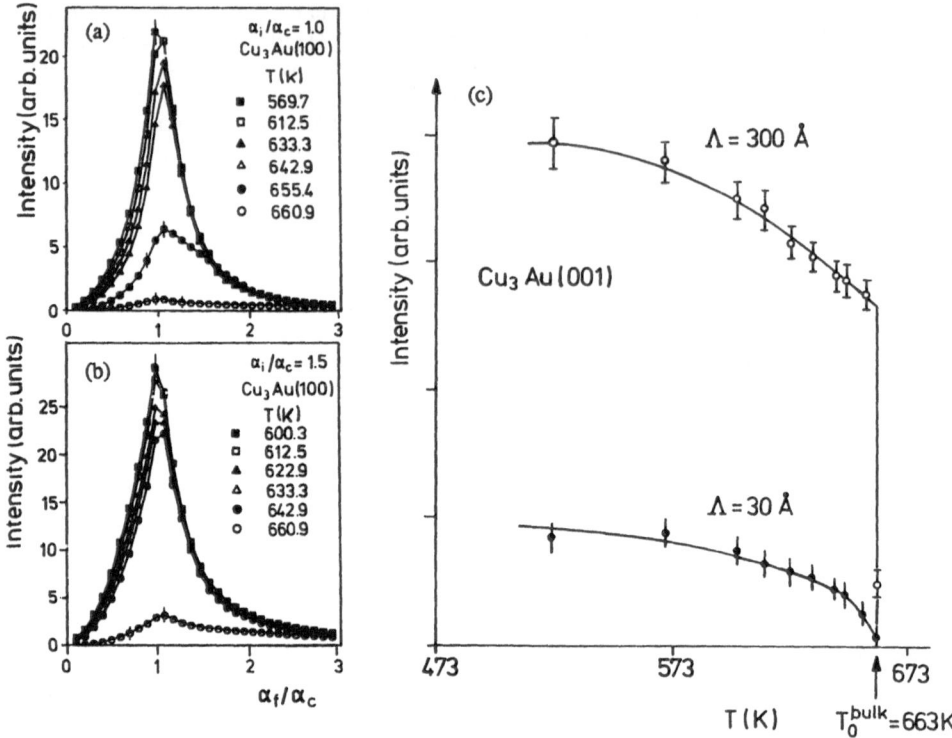

Fig. 5.5a-c. Temperature dependence of the (100) superlattice intensity. **(a),(b)** α_f-profiles associated with $\alpha_i/\alpha_c = 1.0$ and 1.5; **(c)** evanescent intensity I_Λ associated with two scattering depths $\Lambda = 30\,\text{Å}$ and 300 Å

The essence of the experimental observations by Dosch et al. [1988,1991a] is summarized in Fig. 5.5a,b which shows α_f-profiles of the (100) Bragg intensity in thermal equilibrium at two incidence angles $\alpha_i/\alpha_c = 1.0$ and 1.5 for a selection of temperatures below the bulk transition temperature $T_0 = 663$ K. Mere inspection of the raw data shows quite convincingly that the evanescent superlattice intensity disappears in a completely different way in the different α_f-regimes: For $\alpha_f/\alpha_c > 1$ the intensity remains virtually unchanged in the entire temperature range until shortly before the transition, where it drops to zero. On the other hand, for $\alpha_f/\alpha_c \leq 1$, we identify a continuous change of the intensity with each temperature step which is more pronounced for $\alpha_i/\alpha_c = 1.0$ than for $\alpha_f/\alpha_c = 1.5$. This behaviour is demonstrated for two different settings of $\Lambda = 30$ Å and 300 Å (Fig. 5.5c). Since α_f determines crucially the surface sensitivity, a surface-mediated phenomenon is apparently observed, where the surface induces a continuous change of the LRO in agreement with the conclusions drawn from LEED and SPLEED measurements. The authors find within an uncertainty of 0.5 K that the transition temperature T_0 is independent of the distance z, in other words, there is no new phase transition in the investigated near-surface regime.

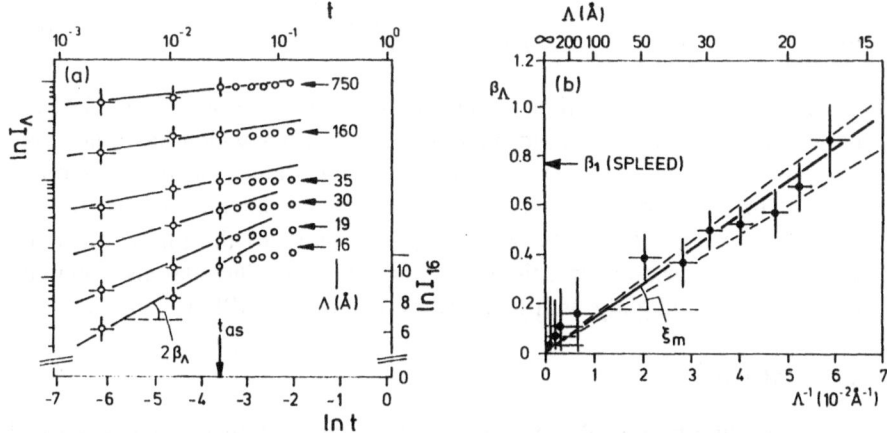

Fig. 5.6. (a) Evanescent superlattice intensity I_Λ versus reduced temperature t on a double logarithmic scale [Dosch et al., 1988]. $2\beta_\Lambda$ denotes the asymptotic powerlaw exponent as observed for $t \leq t_{as}$. (b) Asymptotic powerlaw exponent β_Λ versus Λ^{-1} [Dosch et al., 1988]; ξ_m denotes the slope, β_1 (SPLEED) the powerlaw exponent observed with SPLEED [Alvarado et al., 1987]

For a quantitative analysis of the observed evanescent Bragg scattering the data $I_\Lambda(t)$ associated with various scattering depths Λ (between 16 Å and 750 Å) are plotted versus the reduced temperature t (Fig. 5.6a). The intensities $I_\Lambda(t)$ exhibit a general trend: Starting with a mild temperature dependence far away from T_0 they enter into a marked T-dependence close to T_0 which ends up in an asymptotic power-law behaviour as indicated by the full lines with slope $2\beta_\Lambda$. The quantity β_Λ turns out to depend linearly upon Λ^{-1}, $\beta_\Lambda = \xi_m/\Lambda$ (Fig. 5.6b), thus,

$$I_\Lambda(t) \propto |t|^{2\beta_\Lambda} = |t|^{2\xi_m/\Lambda} \tag{5.6}$$

with (according to Fig. 5.6b)

$$\xi_m = \left(13.1^{+1.1}_{-2.5}\right) \text{Å} \quad . \tag{5.7}$$

It is interesting to note that the experimental value $\beta_1 = 0.77$ as deduced from SPLEED is just comparable to the β_Λ values for small values of Λ (see arrow in Fig. 5.6b), whereas for large Λ the exponent β_Λ tends to zero which is typical for first-order phase transitions.

What is the thermodynamic significance of the scattering depth-dependent powerlaw behaviour (5.6) of the superlattice intensity and of the microscopic length ξ_m (5.7)? In order to answer this question we consider an *"ordinary"* system[3]. Then the mean field solution of $m(z, t)$ assumes the asymptotic form as shown in Fig. 5.3b [Lipowsky and Speth, 1983] which we write for convenience in the form

---

[3] Here, "ordinary" means that the bulk disorder is driven by the surface or that the surface has a tendency to disorder rather than to order.

$$m(z,t) = m_b \cdot \begin{cases} \frac{1}{2}\,e^{[z-L(t)]/\xi_\perp} & \text{for } 0 \le z \le L(t) \\ 1 - \frac{1}{2}\,e^{-[z-L(t)]/\xi_\perp} & \text{for } z \ge L(t) \end{cases} \tag{5.8}$$

where $L(t)$ denotes the position of the interface which is given by (5.4) and $\xi_\perp$ its width ("roughness") due to interfacial fluctuations which have, however, a very smooth temperature dependence: For small values of L the interface is assumed to be smooth, because it experiences the potential of the ordered substrate, accordingly $\xi_\perp \simeq \xi_b = O(a_\perp)$, very close to the phase transition, where L is of mesoscopic size, the beginning interfacial fluctuations eventually tend to increase $\xi_\perp$ slightly with temperature [Lipowsky, 1987],

$$\xi_\perp \propto \xi_b \left(\ln |1/t| \right)^{1/2} \quad . \tag{5.9}$$

The continuum limit of the evanescent Bragg intensity (4.12) associated with this OP profile can be deduced analytically for arbitrary grazing angles (using B.3 in Appendix B). Here we give the result for the superlattice amplitude $F_{SL}(Q_z', t)$ in the case, where the interface is well defined on the scale of Λ, i.e., $\varepsilon_\perp \equiv \xi_\perp/\Lambda < 1$:

$$F_{SL}(Q_z', t) = -m_b \left(\frac{e^{-iQ_z' L(t)}}{\Lambda i Q_z'} + \frac{e^{-L(t)/\xi_\perp}}{2\Lambda i P_z'} \right) \cdot (f_{Cu} - f_{Au}) \tag{5.10}$$

with

$$P_z' \equiv \kappa_z' + i/\xi_\perp \quad . \tag{5.11}$$

Dosch and co-workers [1991a] have recently performed the lattice summation based on (2.37) and (4.12) to obtain $I_{SL}(Q', t)$, resulting in

$$I_{SL}(Q', t) \propto |t|^{2\xi_d/\Lambda} \left[1 - \varepsilon_\perp |1/t|^{1+\varepsilon_\perp} + O(\varepsilon_\perp^2) \right] S_B(Q') \quad , \tag{5.12}$$

where $S_B(Q')$ is given by (2.37). The temperature dependence of the evanescent superlattice intensity associated with a fixed scattering depth Λ, $I_\Lambda(t) \propto |t|^{2\xi_d/\Lambda}$, exhibits a powerlaw behaviour (5.12) of evanescent Bragg scattering which apparently is a direct indicator of a wetting transition governed by short ranged interactions (for a rigorous derivation of the power-law behaviour see (2.43,44) and the subsequent comments on p. 20). By comparison of (5.6) with (5.4,12) we can identify the microscopic length ξ_m as the growth amplitude ξ_d of the disordered phase and get our final experimental result

$$L(t) = \left(13.1^{+1.1}_{-2.5} \right) \text{Å} \ln |1/t| \quad . \tag{5.13}$$

According to (5.12) a nonzero $\xi_\perp$ is observable as a deviation of $I_\Lambda(t)$ from the simple powerlaw behaviour in the asymptotic regime for small values of t. What has been observed in this grazing-angle scattering experiment is apparently the phenomenon that prior to the order-disorder transition the superlattice structure "melts" already at the surface within a certain film whose thickness smoothly increases upon approaching T_0. Because this phenomenon has a close relationship to *surface melting* (Sect. 5.2), we may call this surface effect *"superlattice surface melting"*. Since the driving bulk transition is first order, ξ_d is expected to remain finite on a microscopic range throughout the entire temperature regime, therefore, the observed value of approx. $3a_\perp$ is quite reasonable. The experimental results by Dosch and co-workers may be compared with the Monte–Carlo calculations by Kroll and Gompper [1987] who also find a logarithmic growth of the wetting layer with $\xi_d \simeq 7.5$ Å. In view of the uncertainties in the experimental value and of the open parameters in the Monte–Carlo simulations this is a quite reasonable agreement between theory and experiment. Note that the Monte–Carlo calculations do in particular not account for any surface-induced changes in the coupling constants which may have a significant effect on the value of ξ_d. This may also be the reason for the large discrepancy between the calculated value of β_1 and the experimental one as discussed above, very likely, however, another phenomenon may play a more crucial role: Buck et al. [1983] have shown that the Cu_3Au (001) surface favours the ideally stoichiometric composition, thus, when the bulk composition is slightly off the ideal stoichiometry, the top layer may exhibit a higher transition temperature than the bulk. Interestingly an early LEED measurement (which has not enough been noticed) already reported indications of such a persisting surface LRO at the (001) surface of $Cu_{72}Au_{28}$ [Højlund Nielsen, 1973]. This has recently been confirmed by Liang [1991] using x-rays: Above T_0 associated with $Cu_{3-y}Au_{1+y}$ a surface LRO peak remains which decays softly upon further heating and exhibits hysteresis effects. This observation is also very similar to the one reported at the $Fe_{3-y}Al_{1+y}$ ($1\bar{1}0$) surface by Mailänder et al. [1990a] (see Fig. 4.11). Apparently one observes in both alloys that the near-surface LRO decays faster than the bulk LRO upon approaching T_c or T_0 which is called surface-induced disorder and, parallel to that, a surface segregation-mediated LRO which survives the bulk phase transition and could thus be called surface-induced order. My suggestion is to decompose the surface LRO in such binary alloys into two parts: a "semi-infinite" LRO which may or may not exhibit SID and the associated critical phenomena and a surface segregation-mediated "2d-like" LRO originating from an ideal surface or near-surface layer. The latter may disorder at a temperature $T \geq T_0$ and should exhibit a "2d-like" character. A first theoretical consideration of surface effects of this kind has been reported by Helbing et al. [1990] for a DO_3 structure which undergoes a discontinuous order-disorder transition. Here one finds that various ordered structures may appear at the surface depending on the strengths of the applied bulk and surface fields. Up to now it is unclear, how pronounced these effects are in congruent Cu_3Au. Recently Liang [1991] and Reichert et al. [1992] also showed in an x-ray scattering study of the Cu_3Au (001) surface that there can be a persisting surface LRO above T_0, whereby in the first study the composition is not exactly known, in the latter case one finds $c_{Cu}/c_{Au} = 3.015$ [Dosch et al., 1991a]. It is by no means understood though which parameters, aside from surface segregation, surface miscut, surface adsorbates and the thermal treatment, control the occurence and the extent of this surface effect. A recent LEIS study of the Cu_3Au(001) surface at least indicates that oxygen chemisorption has a crucial influence on the surface composition which enters the surface OP [Nakanishi et al., 1992]. The current understanding of the Cu_3Au (111) surface is even more puzzling: In a first x-ray study one apparently has found persisting surface LRO above T_0 which was called SIO [Zhu et al., 1988]. This findings have not been confirmed by a later, more detailed, x-ray experiment at the same sample (described below), where instead strong surface-induced disorder has been detected [Zhu et al., 1990]. The complication here is that the (111) surface does not terminate ideally but with a slight Au enrichment [McDavid and Fain, 1975], as a consequence, the temperature dependence of $m(z, T)$ (5.5) is not only governed by the ordering densities $r_{\alpha,\beta}(z, T)$, but also by the non-ordering densities $c_{Cu,Au}(z, T)$. This circumstance has to be taken into account properly in future studies.

As discussed above the interaction in Cu_3Au is complicated and certainly of long ranging character, thus, it appears at first sight surprising that the observed wetting phenomenon exhibits a logarithmic growth law (5.13) which is typical for short ranged interactions, $\sim \exp(-L/\xi_d)$. However, we will see in the context of *surface melting* that the long ranging part of the interactions which enters the wetting scenario scales with $(\varrho_{ord} - \varrho_{dis})/L^2$, where ϱ_{ord} and ϱ_{dis} are the densities of the ordered and disordered phase. In the case of Cu_3Au the atoms experience static displacements from their ideal lattice positions (*"size effect"*), when the alloy is in the disordered phase. As a consequence, the associated lattice parameter is slightly enlarged by the superposition of the long ranging tails of these lattice distortions. This subtle effect has already been observed in Cu_3Au by early dilatometer measurements [Nix and McNair, 1941]. From precise measurements of the lattice parameter change across the phase transition by x-ray diffraction one gets $(a_{ord} - a_{dis})/a_{ord} = -1 \times 10^{-3}$ at T_0 [Keating and Warren, 1951], i.e. $(\varrho_{ord} - \varrho_{dis}/\varrho_{ord})$ is as small as 3×10^{-3}. Thus, $(\varrho_{ord} - \varrho_{dis})/L^2$ should overcome $\exp(-L/\xi_d)$ only for large values of L very close to T_0 (we will touch upon this point again at the end of Sect. 5.2.3).

Another aspect of this surface wetting transition is of interest: At the surface of an ordered binary alloy the OP is forced to jump from $m = 1$ for $z > 0$ to $m = 0$ for $z < 0$. It is the task of the intercalated wetting layer to help the system reach the state $m = 0$ at $z < 0$. How about internal interfaces? Of particular interest are antiphase boundaries which are generally present in ordered binary alloys. Across an antiphase boundary the OP may have to undergo an even more drastic change, namely from $m = 1$ to $m = -1$, thus, from a simple energy consideration one would also expect that domain walls of an ordered alloy may become wet when the order-disorder temperature is approached. First indications of such a *"domain wall wetting"* have indeed been observed by recent transmission electron microscopies in ordered $CoPt_3$ [Leroux et al., 1990a,b] and ordered $Cu_{83}Pd_{17}$ [Ricolleau et al., 1990]. In the latter system a logarithmic growth of the width of the antiphase boundary has been found as well. I also refer to the computer simulation by Broughton and Gilmer [1983], where *grain boundary melting* processes have been detected. A related phenomenon has been

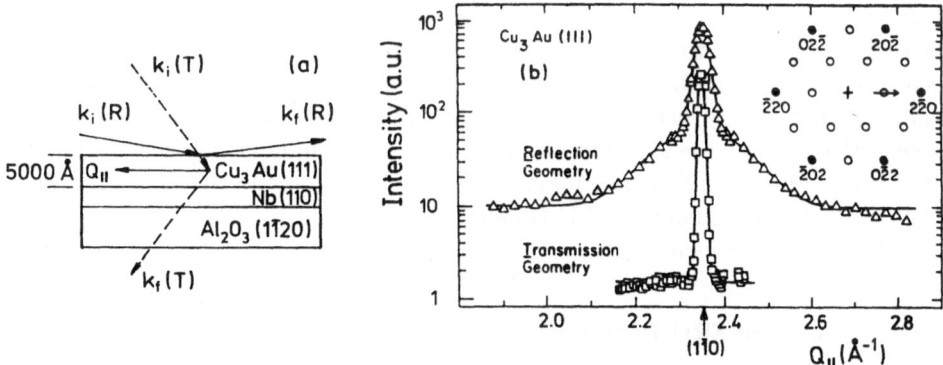

Fig. 5.7a,b. X-ray experiment on Cu₃Au(111) for $T = T_0 - 20\,\text{K}$ [Zhu et al., 1990]; (a) sample (see main text) and scattering geometry: dashed wavevectors are associated with the transmission geometry, the full wavevectors with the GID geometry; (b) radial scans of the $(1\bar{1}0)$ superlattice intensity in transmission and GID geometry. The inset shows a part of the reciprocal lattice with filled circles representing the fundamental reflections and open circles the superstructure reflections

reported by Rabkin et al. [1991]: They apparently observed the wetting of the 43°[100] tilt grain boundary in a $Si_{0.95}Fe_{0.05}$ bicrystal by Sn and Zn.

X-ray measurements of the OP near the phase transition in Cu₃Au have also been performed at the (111) surface [Zhu et al., 1990]. In this case the sample was 5000 Å thick and grown epitaxially on an Al_2O_3 ($1\bar{1}20$) substrate with a Nb (110) buffer layer (Fig. 5.7a): It turns out that also the (111) surface favours the disordered phase as demonstrated in a very direct way: The (110) in-plane superlattice intensity profile was measured in transmission geometry (providing the bulk reference profile) and in the surface-sensitive GID geometry with $Q_z = 0.04 \times (2\pi/a_\perp)$ at $T = T_0 - 20\,\text{K}$. Note that the critical vacuum momentum transfer Q_z^c for total external reflection is $Q_c = 0.023 \times (2\pi/a_\perp)$. While the bulk experiment shows a strong superlattice reflection with almost negligible diffuse scattering, the surface-sensitive experiment discloses a strong broad component originating from near-surface disorder and a relatively weak LRO component (Fig. 5.7b). Interestingly, the in-plane shape of the diffuse scattering has a Gaussian shape and thus suggests that the disordered phase consists of regimes with 25 Å size which are called *heterophase fluctuations* of the *disordered phase*. In the bulk one only finds heterophase fluctuations of the *ordered phase* above T_0 (see e.g. Chen et al. [1977]). We will discuss the associated evanescent structure factor later at the end of Sect. 5.1.3. From MC calculations one finds a strong evidence that the (111) surface should undergo a SID transition independent of the degree of surface segregation [Gompper and Kroll, 1988]. On the other hand it has been predicted that in the case of the $Cu_{50}Au_{50}$ alloy (tetragonal structure) the (010) surface favours the ordered phase, thus, should induce a SIO transition [Schweika et al., 1990]. The occurence of this ordered "CuAuI" structure is shown in the phase diagram (Fig. 5.1a)[4]. The driving mechanism behind this

---

[4] Between the CuAuI structure and the disordered phase is a more complex orthorhombic ordered structure denoted CuAuII.

surface effect is the circumstance that the introduction of the (010) surface removes some of the frustated interactions, therefore the $Cu_{50}Au_{50}$ is more stable close to this surface than in the bulk. This should be readily observable by a grazing angle x-ray scattering experiment. On the experimental side unambiguous surface-induced ordering has so far been observed in liquid crystals [Ocko et al., 1987]. One should also mention a careful ion microscopy study of a Ni_4Mo alloy tip by Kingetsu et al. [1981] who found indications that the (200) and (111) facets favour the ordered phase.

5.1.2 Evanescent SRO Diffuse Scattering

At metal surfaces the pair interaction potential is expected to be changed due to surface-induced changes in the electron density and in the electronic structures (see e.g. Smoluchowski [1944]). Whether or not a given surface of a binary alloy favours order or disorder depends crucially upon the surface interaction energies (J_1) between the atoms (see Lipowsky and Speth [1983]). The experimental measurement of the pair potential V_{lmn} between two surface atoms at r_0 and r_{lmn} could provide a most valuable input into theoretical calculations of surface phase diagrams[5]. From the bulk of an AB alloy system this information is obtained by measuring the SRO diffuse scattering which is observed in the high temperature phase and determined by the Fourier–transform of the pair correlation function

$$g_{lmn} \equiv \langle \tau_{lmn} \tau_0 \rangle - \langle \tau_{lmn} \rangle \langle \tau_0 \rangle \quad . \tag{5.14}$$

By introducing the *Cowley–Warren* SRO parameters

$$\alpha_{lmn} \equiv 4c_A c_B g_{lmn} = 1 - \frac{P_A^{lmn}}{c_A} = 1 - \frac{P_B^{lmn}}{c_B} \tag{5.15}$$

the bulk SRO intensity takes on the simple form

$$I_{SRO}(Q) \propto N c_A c_B \, r_e^2 \alpha(Q) \quad , \tag{5.16}$$

where $\alpha(Q)$ is the Fourier–transform of α_{lmn}. The experimental determination of $I_{SRO}(Q)$ gives access to the SRO parameters α_{lmn}. These SRO diffuse intensities have been measured in Cu_3Au and analyzed with very high perfection by Cohen and co-workers (see Bardhan and Cohen [1976]; Chen et al. [1977]; Butler and Cohen [1989]). In (5.15) P_A^{lmn} $\left(P_B^{lmn} \right)$ is the conditional probability to find an A– (B–)atom as a r_{lmn}-neighbour of a B– (A–)atom[6]. From (5.16) one can deduce the pair interaction potential in a traditional way via the *Clapp–Moss formula* [Clapp and Moss, 1968; Moss and Clapp, 1968]

$$\alpha(Q) = \frac{1}{1 + 4c_A c_B \, V(Q)/k_B T} \quad , \tag{5.17}$$

---

[5] The indices lmn are integer numbers denoting the atomic positions in cartesian coordinates.
[6] Conventionally r_{lmn} is measured in units of $a_0/2$.

where $V(Q)$ is the Fourier–transform of the pair interaction potential. It should be noted, however, that (5.17) accounts for $V(Q)/k_B T$ only up to linear order and is thus reliable only for small values of $V(Q)/k_B T$, i.e., for high enough temperatures and/or not too strong interactions. For this reason more modern methods, like the inverse Monte–Carlo method [Gerold and Kern, 1987], have been developed nowadays.

With the scheme of evanescent scattering it appears possible to extract the equivalent information from the near-surface region of the binary alloy. A first rigorous treatment of the necessary modifications of the bulk scattering theory has been given by Kroll and Wagner [1990]: Due to truncation of the translational invariance at the surface the *Cowley–Warren* SRO parameters become z-dependent close to $z = 0$, i.e. $\alpha_{lmn} \rightarrow \alpha_{lm}(z_n, z_{n'})$, consequently, the near-surface SRO intensity reads

$$
I_{\text{SRO}}(Q_\|, Q'_z) \propto N_S c_A c_B \, r_e^2 \big(T_i T_f^*\big)^2 \big(f_A - f_B\big)^2
$$
$$
\times \sum_n \sum_{n'} \left(\sum_l \sum_m \alpha_{lm}(z_n, z_{n'}) \cos\big(Q_\|^x r_l\big) \cos\big(Q_\|^y r_m\big) \right)
$$
$$
\times e^{-i\big(Q'_z z_n - Q_z^{*'} z_{n'}\big)} \, . \tag{5.18}
$$

By way of example Kroll and Wagner [1990] considered the ideal (i.e. nonsegregated) Cu$_3$Au (111) surface and calculated $I_{\text{SRO}}(Q_\|, Q'_z)$ along $Q_\| = Q_\|(1\bar{1}0)$ for various sets of surface interaction parameters and assuming $T = 1.1 T_0$, grazing angles $\alpha_i = \alpha_f \simeq \alpha_c/2$ and an x-ray wavelength of $\lambda = 1.3$ Å[7] giving $Q'_z = 2.6 \times 10^{-3}$ Å^{-1} + i/18 Å. In the case of diffuse scattering the small real part Re$\{Q'_z\}$ which originates from photoabsorption processes can be neglected compared to the planar momentum transfer $Q_\|$. The full curve in Fig. 5.8 is the

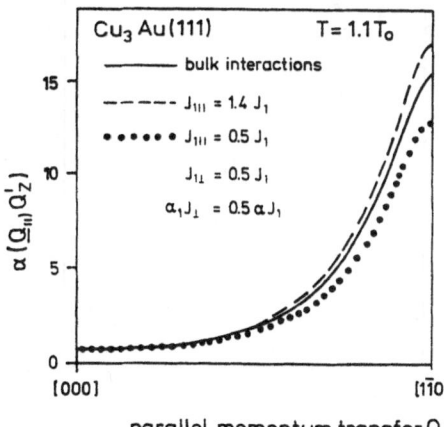

Fig. 5.8. SRO diffuse scattering at the Cu$_3$Au (111) surface as calculated for various surface interaction parameters (see text) [Kroll and Wagner, 1990]

parallel momentum transfer $Q_\|$

---

[7] In a real experiment $\lambda = 1.3$ Å cannot be used, because this would create an untolerably strong Cu fluorescence background.

prediction for $I_{SRO}(Q_\parallel, Q_z')$ for the ideal surface termination, where the surface interactions are not altered, the dashed curve results from an enhancement of the surface in-plane NN interaction ($J_{1\parallel}$) by 40 % which, however, overestimates the pair correlations considerably. When one assumes on the other hand that $J_{1\parallel}$, the out-of-plane NN interaction ($J_{1\perp}$) and in addition the surface out-of-plane NNN interaction ($-\alpha_1 J_\perp$) are decreased by 50 %, then the dotted curve results. One of the messages from this theoretical study is certainly that the effect of surface-modified interaction energies upon the evanescent SRO diffuse scattering is comparatively small, even though the surface sensitivity of evanescent x-rays in Cu₃Au is, with $\Lambda_{min} = 15$ Å (see A.5), very high.

A first experimental study of surface SRO diffuse scattering has been performed by Dosch and co-workers [1991c] on the Cu₃Au (001) surface. The incentive for this experimental effort, however, was not primarily to measure surface-modified interactions, but to get some clues to understand the observation of the surface-induced layered structure above T_0 (Fig. 5.2c) whose origin has been rather unclear. The experiment has been carried out at the wiggler beam line of HASYLAB using a wavelength of $\lambda = 1.72$ Å (for reasons mentioned above). Note by the way that the associated dispersion corrections are very large, $\Delta f_1^{Cu} = -2$ and $\Delta f_1^{Au} = -5$. For two settings of the incidence angle $\alpha_i/\alpha_c = 0.7$ and 1.3 the α_f-profiles of the evanescent SRO diffuse scattering has been recorded along $Q_\parallel = Q_\parallel (100)$ and $Q_\parallel (110)$. The sample temperature was kept throughout the experiment at $T/T_0 = 1.10$. Figure 5.9a,b shows the α_f-integrated diffuse

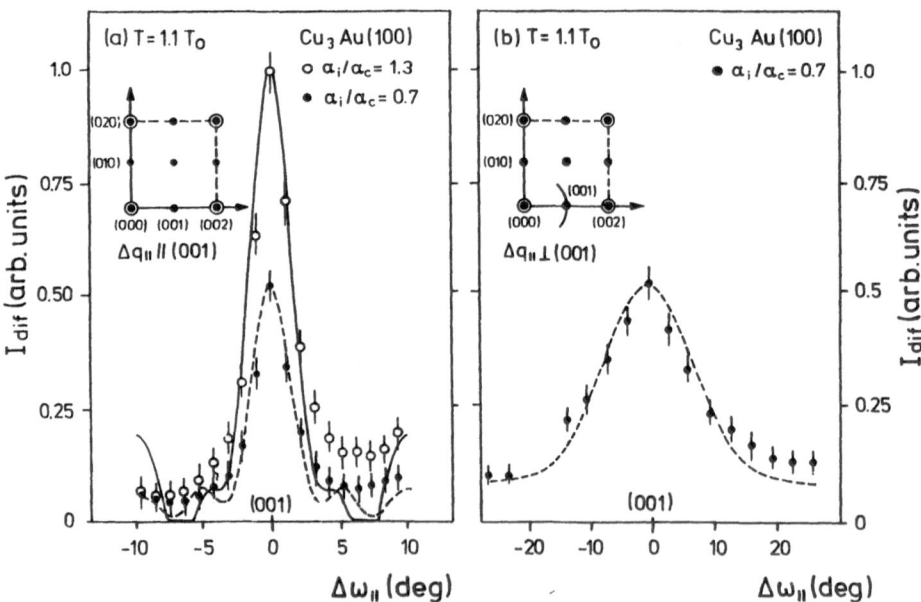

Fig. 5.9a,b. (100) SRO diffuse scattering at the Cu₃Au (001) surface as observed at $T = 1.1\,T_0$ [Dosch et al., 1991c]. (a) Radial scans (see inset) for $\alpha_i/\alpha_c = 0.7$ and 1.2, (b) angular scan (see inset) for $\alpha_i/\alpha_c = 0.7$. The curves are theoretical models as explained in the text

spectra as taken at $\alpha_i/\alpha_c = 0.7$ (penetration depth $\ell_i \simeq 40\,\text{Å}$) and $\alpha_i/\alpha_c = 1.3$ ($\ell_i = 350\,\text{Å}$) along the indicated paths for $Q_{\parallel}$. A rigorous analysis of the observed evanescent SRO scattering would require the proper accounting for the whole set $\{\alpha_{lm}(z_n, z_{n'})\}$ in (5.18). In order to reduce the amount of free parameters Dosch et al. [1991c] replaced $\alpha_{lm}(z_n, z_{n'})$ by a new set $\{\alpha_{lmn}\}_{\ell_i}$ averaged over the illuminated surface slab $\delta F_{\parallel} \cdot \ell_i$ ($\delta F_{\parallel}$ is given by (2.47) and ℓ_i by (2.13)), then, one would expect that the bulk SRO parameter should do a good job for $\ell_i = 350\,\text{Å}$: The full line in Fig. 5.9a is in fact calculated using the bulk SRO parameters as given by Bardhan and Cohen [1976] for $T = 1.12\,T_0$ and provides (for the actual purpose) a fair enough agreement with the experimental data associated with $\ell_i = 350\,\text{Å}$. On the other hand the surface sensitive measurements can be reproduced by the fit (dashed lines in Fig. 5.9a,b) when the "ferromagnetic" NNN correlation α_{200} are enhanced from the bulk *Bardhan–Cohen* value $\alpha_{200}^b = 0.138$ to the large near-surface value $\alpha_{200}^s \simeq 0.6 - 0.8$. The conjecture now is that these surface-enhanced "ferromagnetic" correlations in the SRO produce the observed layered near-surface structure. This would very naturally explain the temperature dependence shown in Fig. 5.2c as the softly decaying near-surface SRO.

With this first experiment it could be demonstrated that the observation of near-surface SRO diffuse scattering is feasible, however, one needs the intense x-ray beam provided by synchrotron wigglers or undulators. From this present experimental experience one can safely say that similar experiments with neutrons, as sometimes suggested, are clearly unrealistic, mainly for two reasons, the limited surface sensitivity of evanescent neutron waves (typically $\geq 100\,\text{Å}$) and the much too low neutron flux available (see A.5 and A.6).

5.1.3 Time Resolved Evanescent Bragg Diffraction

The dynamics of a first order phase transition is characterized by a nucleation process. After a temperature drop from the high temperature into the low temperature phase the ordered phase condensates at natural inhomogeneities, then the radius R of the ordered regions grows with time τ, thereby following a powerlaw

$$R \propto \tau^{-x} \quad , \tag{5.19}$$

where the exponent $x < 0$ depends on details of the cluster growth mechanism. Quite generally one has to distinguish between phase separation processes (*"conserved order parameter"*) and order-disorder processes, where the OP is not conserved. For a discussion of this matter I refer the reader to the excellent review articles by Hohenberg and Halperin [1977], Gunton et al. [1983] and Binder and Heerman [1985]. The equation of motion for the Fourier component m_Q of the OP can be written in the general form (see e.g. Furukawa [1983])

$$\frac{dm_Q}{d\tau} = -D \cdot Q^{\theta} \mu_Q(\tau) \tag{5.20}$$

with μ_Q as the Fourier component of the driving chemical potential and D being the relevant diffusion coefficient which is e.g. $D = D_0/R$ for particle transfer through the interface of the ordered cluster ("*surface diffusion*"). Furthermore $\theta = 0$ for the case of a nonconserved and $= 2$ for a conserved OP. After setting $H_Q \equiv \langle |\mu_Q m_Q^*| \rangle \propto R^\phi$ for the R-scaling property of the energy H_Q of the process under consideration one obtains from rather simple dimensional reasoning the relation (see Furukawa [1983])

$$x = 1/(\phi - d - \theta - 1) \tag{5.21}$$

for the powerlaw behaviour (5.19). The introduced exponent ϕ can be used to specify in a convenient way the nucleation process: In particular, $H_Q = O(k_B T)$, i.e. $\phi = 0$, for the *diffusion-reaction model* of Binder and Stauffer [1974] and $H_Q \propto \gamma \cdot R^{d-1}$, i.e. $\phi = d - 1$, if the surface energy γ is the driving force. In the case of order-disorder transitions $\theta = 0$, thus, one expects $x = -1/4$ in the early nucleation stage, where the nucleus grows via particle transfer through the interface within the *Binder–Stauffer mechanism*, and later $x = -1/2$, when the growth of the nuclei is driven by the surface energy ("grain coarsening regime"). As will be discussed below, the temporal growth of the radius of the nuclei can most directly by observed via the halfwidth $q_2(r)$ of the superlattice peak. Neutron measurements of this kind revealed this crossover from $x = -1/4$ to $x = -1/2$ in the $L1_2$ ordered alloy Ni_3Mn, which undergoes a discontinuous order-disorder transition at $T_0 = 783$ K [Katano and Iizumi, 1988]. The time dependence of the Cu_3Au bulk ordering has been studied experimentally by various groups (see e.g. Nagy and Nagy [1962]; Noda et al. [1984]; Nagler et al. [1988]). The bulk superlattice intensity can be fitted in a first approximation by [Nishihara et al., 1982]

$$I(\tau) = I_\infty(1 - e^{-\tau/\tau_K}) \quad , \tag{5.22}$$

where τ_K measures the escape rate from the disordered in the ordered state.

Here I will address the question, how the presence of a free surface which favours the disordered phase will modify the relaxation behaviour, in particular the time constant τ_K and the characteristic exponent x (5.21). Before discussing some time resolved evanescent x-ray scattering experiments I refer the reader to theoretical studies of the temporal growth of wetting layers that are reviewed by de Gennes [1985]. Recently, Lipowsky [1985] considered the influence of short and long range interactions on the growth characteristics of wetting layers when the temperature is raised from $T < T_0$ to $T = T_0$. First experimental evidence of the time dependence of the surface LRO has meanwhile become available from LEIS and LEED [McRae and Buck, 1990] and x-ray scattering experiments [Zhu et al., 1989; Dosch et al., 1991a; Reichert et al., 1992]. It turns out that the typical relaxation time τ depends sensitively upon the crystal surface and the associated surface segregation phenomena: For $T < T_0$, i.e., *in the presence of an ordered bulk*, the top layers of the (001) surface may order with a time constant down to only a few seconds and thus a factor 10^5 faster than the bulk [McRae and Buck, 1990]. The ideal compositional termination of this surface obviously entails this rapid ordering. At the (110) and (111) surfaces, where a temperature-dependent surface enrichment of Au occurs [McRae and Buck, 1990; McDavid and Fain, 1975], the associated LRO relaxation times of the surface superlattice reflection are found to be comparable to or even longer than the bulk values. Apparently, the establishment of surface LRO requires there, due to surface segregation, a long-ranging transport of the atoms from and to the surface.

This complicated situation at the surface of binary alloys becomes a little more transparent, if one keeps in mind that $m(z)$ is governed by the occupation probabilities $r_y(z)$ $(y = \alpha, \beta)$ and by the actual concentration $c_x(z)$ of the constituent x (here $x = Cu, Au$). If $c_x(z)$ changes in the temperature regime around the order-disorder transition temperature, this affects $m(z)$ and its relaxation behaviour. Only at ideally terminating surfaces, as the (001) surface of Cu_3Au, one has the possibility to study the relaxation process of the ordering density $r_y(z)$.

Depth- and time-resolved grazing angle x-ray scattering experiments have been reported by Zhu et al. [1989] at the Cu_3Au (111) surface and by Dosch et al. [1991a] and Reichert et al. [1992] at the Cu_3Au (001) surface. In the latter experiment time-resolved α_f-profiles of the (100) Bragg scattering were measured during a quench from $T_{init} > T_0$ to $T_{final} \simeq T_0 - 36$ K and depth profiles of the near-surface relaxation behaviour of the surface LRO between $\Lambda = 20$ Å and $\Lambda = 300$ Å were obtained in this way. The results for $\chi_\Lambda(\tau) \equiv I_\Lambda(\tau)/I_\Lambda(\infty)$ for $\Lambda = 20$ Å, 35 Å, $\simeq 100$ Å and 200 Å are summarized in Fig. 5.10a together with the exponential fits and the associated τ_K-values (see inset of Fig. 5.10a) which increase very distinctly with decreasing scattering depth. The important observation is that the LRO associated with large scattering depths outcrops first followed by the surface LRO which has a noticeably larger initial time constant. This near-surface relaxation behaviour may be understood from the equilibrium properties of the surface LRO: The Cu_3Au (001) surface exhibits a socalled "ordinary" behaviour of the surface LRO, i.e., the near surface regime favours disorder and the near surface LRO is driven by the bulk. Thus, any near surface LRO can only develop in the presence of LRO in the bulk of the system, consequently, one would expect that in a temperature quench from the fully disordered state the near surface LRO should always lag temporarily behind the bulk LRO. This conclusion is also in qualitative agreement with the experimental results from the (111) surface [Zhu et al., 1989]. The derived quantities $\Delta\chi_{12}(\tau) \equiv \chi_{\Lambda_1}(\tau) - \chi_{\Lambda_2}(\tau)$ for $\Lambda_1 = 220$ Å and $\Lambda_2 = 20$ Å, 35 Å and 100 Å (Fig. 5.10b) display maxima $\Delta\chi_m(\Lambda_2)$ which become more and more pronounced as the surface sensitivity is increased. τ_{max} ($\simeq 16$ min $- 17$ min) denotes the time after the quench where the bulk has its biggest temporal lead over the lagging near surface region.

We have seen that the relaxation time τ_K shows a dramatic z-dependence, so it is pertinent to ask, whether the nucleation process is also significantly altered by the presence of the surface. The microscopic nature of the growth mechanism can be assessed from the time-dependence of the half width $q_2(\tau) \propto R^{-1}(\tau)$ of the superlattice reflection. The relaxation of the LRO near the surface after a quench from $T = T_0 + 5$ K to $T = T_0 - 10$ K has been measured by Zhu et al. [1990]

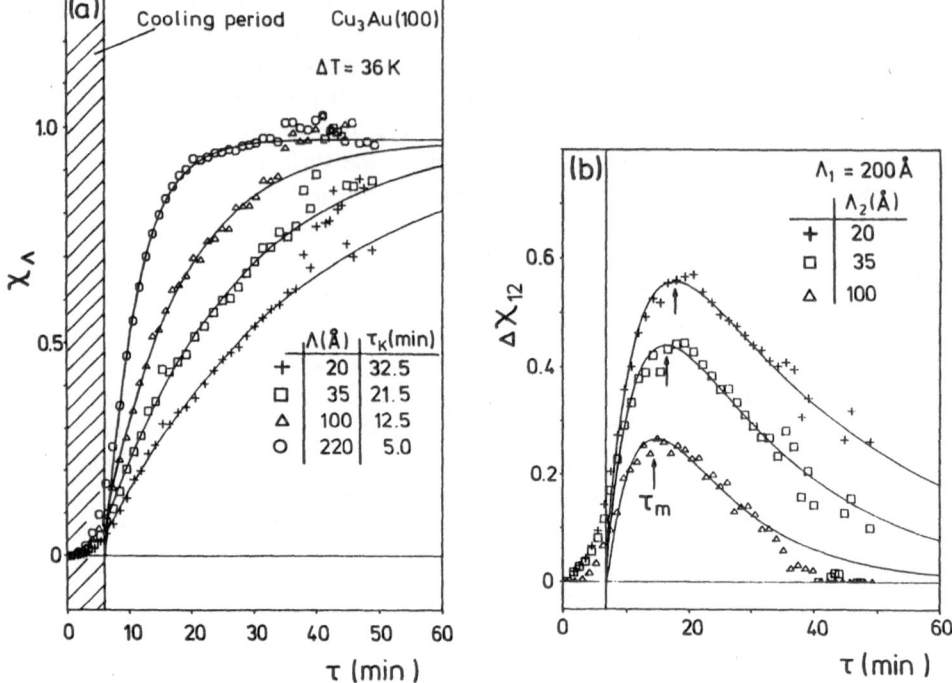

Fig. 5.10a,b. Time dependence of the (100) evanescent superlattice intensity at the Cu_3Au (001) surface after a temperature quench from $T > T_0$ to $T \simeq T_0 - 36$ K. (a) Normalized intensity $\chi_\Lambda \equiv I_\Lambda / I_\infty$ for various values of Λ; (b) quantity $\Delta\chi_{12} \equiv \chi_1 - \chi_2$ for $\chi_1 = \chi_{200}$ and $\chi_2 = \chi_{20}$, χ_{35} and χ_{100}. τ_m denotes the elapsed time when the bulk order parameter has its biggest temporal lead to the near surface order parameter.

on the Cu_3Au (111) surface at a grazing angle $\alpha_{i,f} \leq \alpha_c$ and in transmission geometry (bulk reference) (Fig. 5.11a). The analysis of the observed halfwidth q_2 of the (110) reflection gives

$$q_2^{110}(\tau) \propto \tau^{-0.4} \quad \text{for } \tau \geq 35 \text{ min} \tag{5.23}$$

both for the surface and the bulk measurement. This observation is characteristic for the late stage, where one has a growth driven by the surface energy ($x = -1/2$).

In the time-resolved experiments by Reichert et al.[1992] on the Cu_3Au (001) surface the system was quenched to a final temperature $T \simeq T_0 - 36$K (similar to the bulk experiments in Ni_3Mn by Katano and Iizumi [1988]). For two incidence angles $\alpha_i = 0.63$ and $1.1\,\alpha_c$ the radial distribution of the (100) Bragg scattering has been measured, thereby integrating $\alpha_f = 0.4 - 1.6\,\alpha_c$. Interestingly, here one observes (Fig. 5.11b)

$$q_2^{100}(\tau) \propto \begin{cases} \tau^{-0.22} & \text{for } \tau \leq 50 \text{ min and } \alpha_i = 1.1\,\alpha_c \rightarrow \Lambda = 220 \text{ Å} \\ \tau^{-0.29} & \text{for } \tau \leq 50 \text{ min and } \alpha_i = 0.63\,\alpha_c \rightarrow \Lambda = 30 \text{ Å} \end{cases} \tag{5.24}$$

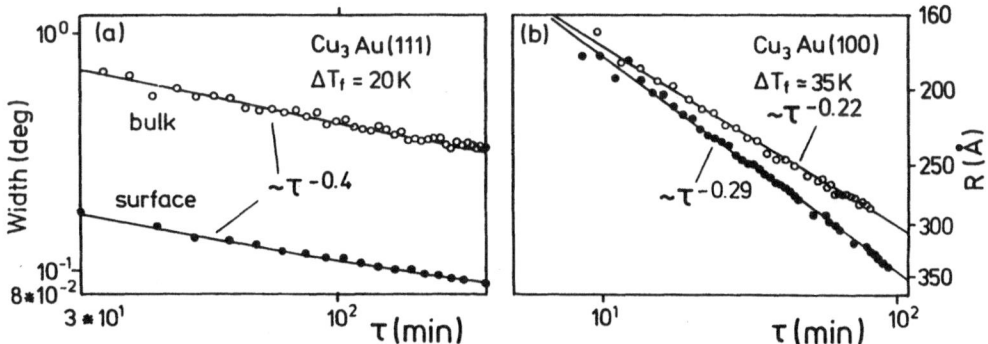

Fig. 5.11a,b. Halfwidth of the superlattices intensity versus time after a temperature quench: (a) at the Cu₃Au (111) surface after a quench to $T = T_0 - 10\,K$ [Zhu et al., 1989], (b) at the Cu₃Au (001) surface after a quench to $T \simeq T_0 - 36\,K$ [Reichert et al., 1991]

The exponent x associated with $\Lambda = 220$ Å is close to $x = -1/4$ characteristic for the Binder–Stauffer mechanism (i.e. $\phi = 0$ in (5.21)) which, however, has never been observed in Cu₃Au bulk experiments. A quite general experimental experience is that the relaxation behaviour in this early stage depends upon the thermal history of the sample. In addition, the comparison of the various quench experiments is not straightforward, because of the different quench conditions: In the case of bulk Cu₃Au one has never applied such a deep quench to $T_0 - 36\,K$ below the spinodal temperature $T_{sp} = T_0 - 24\,K$ [Chen et al., 1977; Gaulin et al., 1991], on the other hand, the similar deep quench in the related system Ni₃Mn also yields an initial relaxation behaviour with $x = -1/4$ in the same time regime ($\tau \leq 60\,min$). Therefore, this result reported by Reichert et al. [1992] is probably a realistic effect and not a scattering artefact as e.g. brought about by the α_f-integration. As the surface sensitivity is enhanced ($\Lambda = 30$ Å) the exponent x grows to $x = -0.29$. Supposed that this new surface effect will be confirmed by future experiments, there would exist no theoretical model up to now which could explain it. When one notes, however, that $x = -1/3$ in the case of phase separation ($\theta = 2$), then one may speculate whether the surface fields, which are responsible for the layered structure above T_0 (Sect. 5.4.3), may also cause the observed change in x in the early nucleation stage.

I want to end this with some remarks on the two complementary ways how the Cu₃Au (001) and (111) surfaces disclose their predilection for the disordered phase:

a) *Non-equilibrium scattering profiles*: In a temperature quench from the disordered state the ordering near the surface is driven by the bulk order as shown schematically in Fig. 5.12a: Shortly after the quench there is a surface regime of a thickness L^*, where no significant ordered clusters are present. This leads to distinct evanescent scattering phenomena: Assuming uncorrelated ordered domains of size $s_b^* \pm \delta s_b$ (in units of $a_\perp$) to be present in deeper regions of the quenched alloy, say starting at L^*, the scattering intensity associated with these ordered domains along Q_z' is given by the expression

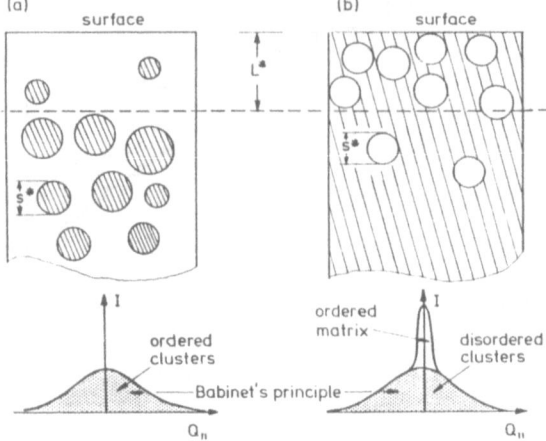

Fig. 5.12. Schematic situation at the surface of a binary alloy (a) shortly after a quench, (b) in thermal equilibrium. L^* denotes the surface regime which exhibits reduced order. The associated scattering signals along $q_{\parallel}$ are indicated (using Babinet's principle); for further explanation see text

$$S_1\left(Q'_z; L^*, s_b^*; \delta\right) = \sum_{q=L^*}^{\infty} \left\langle \left| \sum_{z=q}^{q+s_b} e^{-iQ'_z a_\perp z} \right|^2 \right\rangle_{\delta s_b}$$

$$= N_b S_B(Q'_z) H_1(L^*) \left\langle \left| C\left(-iQ'_z s_b a_\perp\right) \right|^2 \right\rangle_{\delta s_b} \qquad (5.25)$$

with $N_b \equiv \left(1 - e^{-2s_b a_\perp/\Lambda}\right)^{-1} \simeq \Lambda/(2s_b a_\perp)$, $\langle\ \rangle_{\delta s_b}$ as the average over the cluster size distribution,

$$C(x) \equiv 1 - e^x \qquad (5.26)$$

as the "*coherence amplitude*" and

$$H_1(L^*) = e^{-2a_\perp L^*/\Lambda} \quad . \qquad (5.27)$$

b) *Equilibrium scattering profiles*: In the thermally equilibrated ordered state small disordered clusters seem to occur at the surface as illustrated in Fig. 5.12b. According to *Babinet's principle*[8], the scattering intensity $S(Q_{\parallel})$ of these disordered clusters is the same as the scattering from the ordered clusters in the disordered matrix (see schematic intensity distribution at the bottom of Fig. 5.12), the only difference is apparently the depth regime in which these clusters occur. Consequently, the associated evanescent scattering cross sections $S_2\left(Q'_z; L^*, s_b^*; \delta\right)$ should show distinct features which allow one to distinguish it from case (a). Applying *Babinet's principle* we find for the case (b) in Fig. 5.12 that the asso-

---

[8] Two objects $O_1(r)$ and $O_2(r)$ are called complementary when $O_1(r) + O_2(r) = 1$. The Fourier transform of this yields $P_2(q) = \delta(q) - P_1(q)$ for the Fourier amplitudes $P_{1,2}(q)$, thus, except for $q = 0$, $|P_1|^2 = |P_2|^2$.

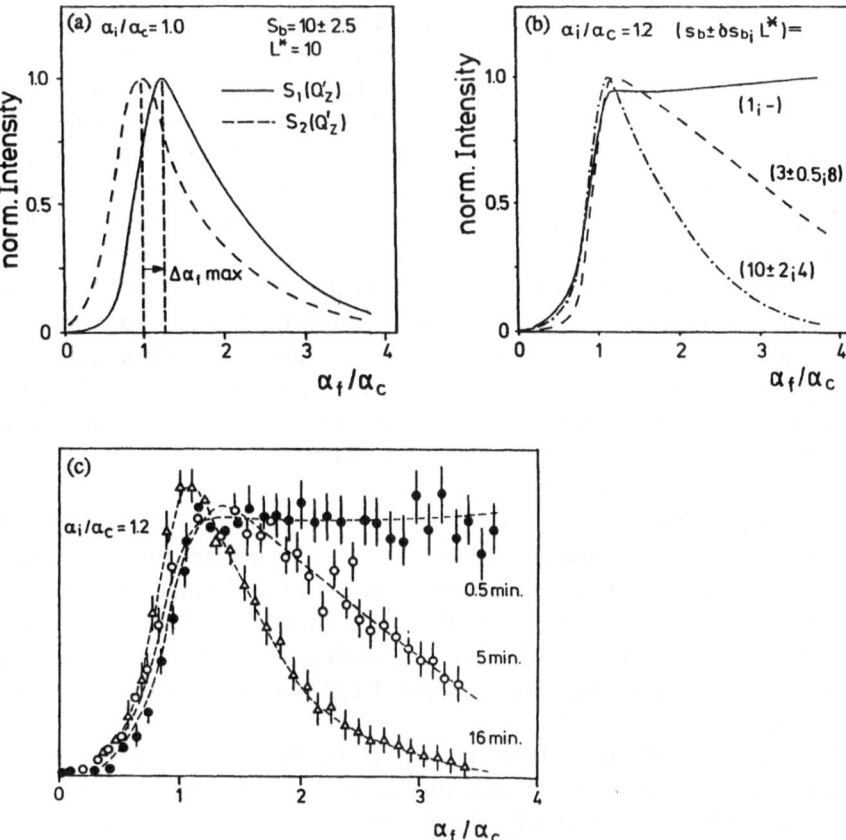

Fig. 5.13a-c. α_f-profiles of small ordered and disordered clusters in Cu₃Au. (a) Scattering intensities associated with the cross sections S_1 and S_2 for clusters of size $s_b = 10 \pm 2.5$ and $L^* = 10$. (b) Short time scenario of the OP relaxation in Cu₃Au simulated with the S_1 scattering cross section for $\alpha_i/\alpha_c = 1.2$ and various parameters ($s_b \pm \delta s_b$; L^*). (c) Observed α_f-scattering profiles at the Cu₃Au (001) surface shortly after a temperature quench [Reichert et al., 1992]

ciated scattering law S_2 is almost identical to S_1 (5.25) with the only exception that the external sum $\sum_{q=L^*}^{\infty}$ has now to be replaced by $\sum_{q=0}^{L^*}$ yielding (again assuming uncorrelated clusters)

$$S_2(Q'_z; L^*, s_b^*; \delta) = N_b S_B(Q'_z) H_2(L^*) \left\langle \left| C(-iQ'_z s_b a_\perp) \right|^2 \right\rangle_{\delta s_b} \qquad (5.28)$$

which differs from S_1 (5.25) by the damping factor

$$H_2(L^*) = 1 - e^{-2a_\perp L^*/\Lambda} \quad . \qquad (5.29)$$

Figure 5.13a shows some theoretical α_f-intensity profiles based on (5.25, 28) using $L^* = 10$ and $s_b^* = 10 \pm 2.5$ calculated for the case of Cu₃Au (001) with $\alpha_i/\alpha_c = 1.0$. The scattering profiles are indeed different from each other in a characteristic way, in particular, the intensity maximum of S_1 is shifted by

$\Delta\alpha_f^{max}$ from $\alpha_f/\alpha_c \simeq 1$ to larger values due to the action of the damping factor $H_1(L^*)$. In Fig. 5.13b the *normalized* theoretical α_f-scattering profiles S_1 (5.25) associated with the short time quenching scenario in Cu$_3$Au (001) are shown for $\alpha_i/\alpha_c = 1.2$, a surface roughness of $\varrho = 8$ Å and for various parameters $(s_b \pm \delta s_b; L^*)$ starting from the incoherent limit $s_b = 1$ to a cluster size of $s_b = 10 \pm 2$. Interestingly, a similar short time behaviour has been reported by Reichert et al. [1992] in the deep quench of Cu$_3$Au (001): Figure 5.13c shows the observed α_f-scattering distributions as observed at three times $\tau = 30\,\text{s}$, $5\,\text{min}$ and $16\,\text{min}$ after the temperature quench from the disordered phase. Although the experimental evidence is very little, its good quantitative agreement with the theoretical curves in Fig.5.13b seems to confirm the proposed surface scenario. A systematic experimental investigation of the scattering cross sections S_1 and S_2 will certainly provide additional helpful information on the microscopic nature of the SID phenomenon in Cu$_3$Au and other alloys.

I add here two short notes:

a) The special scattering law $S_1(Q'_z; 0, s_b; 0) \equiv S^+(Q'_z; s_b)$, i.e. $L^* = 0$, $\delta s_b = 0$, has a particular importance: $S^+(Q'_z; s_b)$ degenerates to $N_b \propto \Lambda/2a_\perp$ for $s_b = 1$, i.e. to the incoherent scattering law (2.38), and to the Bragg scattering law $S_B(Q'_z)$ (2.37) for $s_b \to \infty$. Thus, the introduced coherence amplitude $C(-iQ'_z s_b a_\perp)$ allows the simplest possible analytical tuning from totally incoherent evanescent scattering to evanescent Bragg scattering by one parameter, the coherent size $s_b \in [1, \infty]$.

b) Note also that at the final stage the ordered domains are *coherently* separated by antiphase boundaries (APB) with displacement vectors $r_{APB} = (011)a_0/2$. Thus, in the (001) direction of Q'_z, the phase shift across an APB is $\exp(i\pi)$ and the contribution of antiphase domains of size $l_D a_\perp$ to the square of the superlattice scattering amplitude is accordingly (assuming infinitely small antiphase boundaries)

$$\left|F_{APB}(Q'_z; l_D)\right|^2 = \left|\sum_{\alpha=0}^{\infty} e^{i\alpha\pi} \sum_{n=\alpha l_D}^{(\alpha+1)l_D-1} e^{-iQ'_z a_\perp n}\right|^2$$

$$= \left|F_B(Q'_z) \cdot \tanh\left(iQ'_z a_\perp l_D/2\right)\right|^2 \qquad (5.30)$$

with

$$F_B(Q'_z) = \left(1 - e^{-iQ'_z a_\perp}\right)^{-1} \qquad (5.31)$$

as the Bragg amplitude of the totally coherent state [see (2.37)]. In order to describe the real system, a domain size distribution δl_D has to be included, thus, approximately

$$S_{APB}(Q'_z; l_D; \delta l_D) = \left\langle \left|F_{APB}(Q'_z; l_D)\right|^2 \right\rangle_{\delta l_D} . \qquad (5.32)$$

As it should be, $S_{APB}(Q'_z; l_D; \delta l_D)$ degenerates to the totally coherent case $S_B(Q'_z)$ (2.37) when $l_D \to \infty$.

5.2 Surface Melting

The *solid-liquid* phase transition is the most ubiquitous example of a cooperative phenomenon we know. Though melting (and freezing) plays a vital role in our everyday life as well as in many technologies, it may appear astonishing that there exists no accepted microscopic theory of melting. One has to realize on the other hand that the understanding of the melting process is intimately related to the insight into the atomic structure of a liquid which is still incomplete. The original approaches to attack the problem of the solid-liquid transition started with the crystalline state which allowed a lattice-theoretical treatment, then, the melting was modelled by the generation of vacancies and other defects. Within this scheme the liquid appeared as a *"strongly disordered solid"*, a conception which is wrong as we know today: When a solid starts to melt, the Bragg spots which indicate *long ranged positional order* disappear completely from the diffraction pattern. The liquid structure is thus characterized by a mean density and *short ranged pair correlations*. We may use the set of Fourier components $\{\varrho(G)\}$ of the electron density $\varrho(r)$ as the OP of the melting transition which has thus infinitely many components. The Landau free energy $F(G)$ associated with the Fourier component $\varrho(G)$ of the solid and the Fourier amplitudes $A(G)$ and $B(G)$ is

$$F(G) = A(G)\, \varrho(G)\, \varrho^*(G)$$
$$+ B(G)\, \varrho(G)\, \varrho(G')\, \varrho(G'')\, \delta(G + G' + G'') + \dots \tag{5.33}$$

and allows in particular for a third order term which renders the solid-liquid transition necessarily first order. However, the weak discontinuity in the average density (compared to the liquid-vapour transition) can be regarded as a hint that the order of the melting transition may become continuous in lower dimensions[9]. It further turns out that the ϱ^3-term in (5.33) favours the discontinuous formation of bcc structures, in order to obtain fcc or other structures higher order terms are necessary.

5.2.1 Criteria for Melting and Surface Melting

Some heuristic criteria of melting have emerged which are sometimes elevated to the status of "melting theories", but are actually only empirical rules. Presumably the most famous one is due to Lindemann [1910] (*"Über die Berechnung molekularer Eigenfrequenzen"*) who argued that a solid starts to melt when the mean square amplitude of the thermal vibrations, $\langle u_{\text{th}}^2 \rangle$, reaches a critical value (see also Stroud and Ashcroft [1972]). In fact at T_m, the high-temperature form of $\langle u_{\text{th}}^2 \rangle$ achieves a limiting value $\sigma_L d_{\text{NN}}$, where σ_L seems to depend only on the crystal structure and is e.g. $\sigma_L \simeq 0.2$ for fcc metals (d_{NN} is the nearest neighbour distance).

---

[9] For $d = 2$ there is a continuous melting transition via a new ("hexatic") phase [Kosterlitz and Thouless, 1972; see also Halperin and Nelson, 1978].

In a systematic experimental study of the temperature dependence of the shear modulus C_{44} Varshni [1970] showed that $\sigma_B \equiv C_{44}(T_m)/C_{44}(T = 0) \simeq 0.5$ appears to be universal for fcc metals, an experimental fact which could be used as a melting rule which originally goes back to Born [1939]: In deriving the general conditions for the stability of crystals he proposed (erroneously, as it turned out) that the shear modulus C_{44} disappears at T_m ($\sigma_B = 0$).

In the 50ies it became more and more clear that defects like vacancies, interstitials and dislocations are important in the understanding of the melting process. In particular it was noted that for elements having the same crystal structure the quantity $\sigma_F \equiv E_v/kT_m$ (where E_v is the vacancy formation enthalpy) has essentially the same value. This implies that the vacancy concentration at the melting point, $n_v(T_m) \sim \exp(-\sigma_F)$, is more or less the same for a given crystal structure. This led Frenkel [1955] to conclude that vacancies may play an important role in the melting process. Since $n_v(T_m) = O(10^{-3})$, it is kind of hard to imagine though that this homeopathic amount of vacancies should be the agent which destroys the entire crystal lattice. In fact today one favours more the idea that melting corresponds to a catastrophic proliferation of dislocations (see e.g. Woodruff [1973]; Cotterill et al. [1974]): At a certain critical dislocation density the crystal gets disrupted and melts. In the dislocation production vacancies may be helpful, but don't seem to be the crucial ingredient. Even though one has no reliable experimental evidence for the validity of the dislocation theory, it is one of its merits that it can account for many changes in physical properties upon melting, as for the volume change, the latent heat, the dramatic change in fluidity or the geometric origine of the Lindemann parameter $\sigma_L \simeq 0.2$.

However, all of these theories miss one very crucial point, namely the answer to the question, *where* the crystal actually starts to melt. An intriguing suggestion is that the melting process is sparked at the surface of the solid (see Fig. 5.13). The conjecture that the melt can emerge at the surface of a solid well below the melting point is in fact very old and goes back to Faraday [1860] who investigated some mechanical properties of ice surfaces. Today it is well established experimentally that the melting temperature depends on the size of the solid. Consider the equilibrium condition $\mu_s(p_m, T_m) = \mu_l(p_m, T_m)$ for the coexistence of the solid and the liquid phase (μ_s, μ_l are the associated chemical potentials): Since in a small spherical particle of radius r, the internal pressure p_a ($a = s, l$) is $p_a = p_{ext} + 2\gamma_{av}/r$ (see Herring [1952]), with γ_{sv}, γ_{lv} as the solid-vapour and liquid-vapour interfacial energies, it is quite evident that the above equilibrium condition becomes dependent upon the particle size r and its surface properties (through the interfacial energies γ) and leads in fact to a lowering of T_m with decreasing r. This thermodynamic size effect has been demonstrated very impressively on small Au particles [Buffat and Borel, 1976] and on small particles of Pb and In [Coombes, 1972]. The energetic situation at the surface of a solid apparently plays a crucial role. Also in most of the above melting rules the surface is important: Following the Lindemann rule it is quite evident that any surface-enhanced mean square displacement of the surface atoms would lead to a premelting of the surface, as a matter of fact one even would conclude from the analysis of the melting of alloys [Voronel et al., 1988] that also static surface relaxation effects lead to a lowering of the surface melting temperature. Regarding Frenkel's vacancy theory it was stressed by Stoltze et al. [1988] that the energy involved in creating a vacancy is remarkably low for the Al (110) surface, $E_v = 0.3\,\mathrm{eV}$, therefore the action of vacancy formation close to an open surface could be important in the understanding of the role of the surface for the melting process. Indeed, in a molecular dynamics simulation of the Al (110) surface below the melting point the disordering of the first surface layers seems to coincide with the onset of an enhanced surface vacancy formation [Stoltze et al., 1989].

In order to understand the thermodynamic role of the surface close to the melting temperature of a system we resort to the general wisdom that the liquid state starts to appear when it has a lower free energy than the crystalline solid. When an undercooled quasiliquid layer of thickness L intervenes between the solid-vapour interface, the gain in free energy is (see Fig. 5.14)

$$\Omega^s(L) = L\left(\Omega_l^b - \Omega_s^b\right) + \gamma_{sv}e^{-L/\xi_d} + (\gamma_{sl} + \gamma_{lv})\left(1 - e^{-L/\xi_d}\right) \quad, \qquad (5.34)$$

with Ω_l^b, Ω_s^b as the bulk free energies of the liquid and the solid, respectively. Since the introduced disordered surface layer contains Fourier components of the solid, the solid-vapour surface energy γ_{sv} does not fully disappear, but fades away with the thickening of the film (2nd term), while simultaneously the two new interfaces ("sl" and "lv") slowly emerge (3rd term). In the limit $L/\xi_d \to \infty$

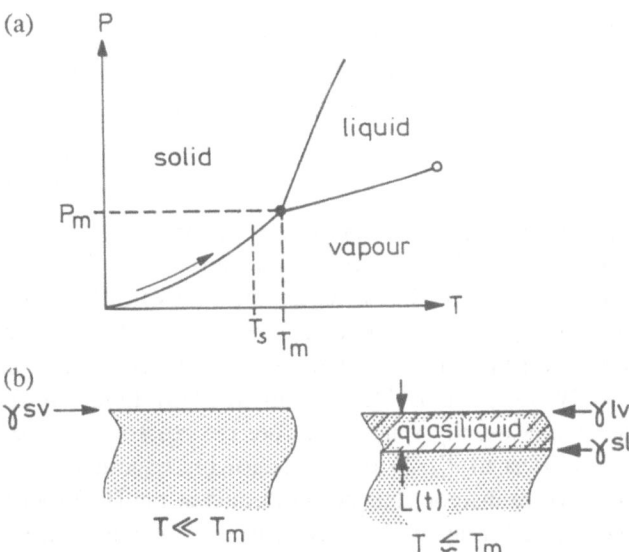

Fig. 5.14. (a) $p - T$-phase diagram of a one component system: T_m is the melting temperature; (b) surface of a solid (left) at a temperature far below T_m (γ_{sv} is the solid-vapour surface energy) and (right) at the onset of surface melting (γ_{sl} is the solid-liquid and γ_{lv} the liquid-vapour interfacial energy

(which corresponds, as we will see in a moment, to $t \to 0$) a thermodynamically stable liquid finally comes out of this and the exponential damping factors in (5.34) are gone. Lipowsky et al. [1983] pointed out that the driving mechanism of this wetting transition can be described by an effective interaction between the two interfaces separated by L. For a small separation short range (SR) interactions dominate giving a term $V(L) = \Delta\gamma \exp(-L/\xi_d)$ to Ω^s [Widom, 1978] with $\Delta\gamma = \gamma_{sv} - (\gamma_{sl} + \gamma_{lv})$. In this picture the effective potential has to be repulsive ($\Delta\gamma > 0$), if surface melting should occur. Then the two interfaces repel each other, until the action of the long ranging van der Waals (vdW) interaction becomes significant and adds $C \varrho_l(\varrho_s - \varrho_l)L^{-2}$ [Lipowsky, 1987] to Ω^s. The sign of the socalled *"Hamaker constant"* $H = C \varrho_l(\varrho_s - \varrho_l)$ depends on the density difference $\varrho_s - \varrho_l$ between the solid and the liquid (the constant C contains details of the underlying vdW interaction). In particular for ice, Ge and Bi which exhibit "regelation" ($\varrho_s - \varrho_l < 0$) this long ranging part of the interface potential is attractive and is thus expected to lead to a sudden stop of the surface melting process when a certain film thickness is achieved (*"blocked surface melting"*; for a recent theoretical study of this effect in ice see Elbaum and Schick, [1991]). Since for a first order phase transition $\Omega_{ls}^b = \Delta S_0^b(T_m - T)$, the effective-interface potential is now

$$\Omega^s(L) = L\Delta S_0^b(T_m - T) + (\gamma_{sl} + \gamma_{lv}) + \Delta\gamma\, e^{-L/\xi_d} + C \varrho_l(\varrho_s - \varrho_l)L^{-2} .$$
$$(5.35)$$

The temperature behaviour of the film thickness can be deduced immediately from (5.35) and the equilibrium condition $\delta\Omega/\delta L = 0$ to be

$$L(T) = \begin{cases} \xi_d \cdot \ln|T_s/(T_m - T)| & \text{for repulsive SR forces } (T < T^*) \\ l \cdot |T_m/(T_m - T)|^\omega & \text{for repulsive vdW forces } (T > T^*) \end{cases}$$
$$(5.36)$$

with

$$T_s \equiv \Delta\gamma/(\xi_d \Delta S_0^b) \qquad\qquad (5.37)$$

which plays the role of the onset temperature for surface melting.

In the temperature regime $T < T^*$ (T^* crossover temperature), where SR interactions govern the phenomena, we find a logarithmic growth of the wet surface layer, for $T > T^*$ and $\varrho_s - \varrho_l > 0$ the action of the vdW forces leads to a crossover at T^* to a powerlaw growth, where $\omega = 1/3$ (1/4) is expected for nonretarded (retarded) vdW interactions[10] [Israelachvili, 1985]. In metals, the crystal-melt density difference is usually small, therefore the crossover temperature T^* is expected to be closest to T_m. With ion shadowing experiments the logarithmic growth law of surface disorder has been detected at the H_2O (0001) surface [Golecki and Jaccard, 1978], at the Pb (110) surface [Frenken and van der Veen, 1985] and recently at the Al (110) surface [Denier van der Gon et al., 1990]. Two points should be addressed here:

---

[10] In (5.35) we have assumed nonretarded vdW forces. In the retarded case L^{-2} has to be replaced by L^{-3}.

a) In (5.35) the expression $L\Delta S_0^b(T_m - T)$ which measures the deviation from bulk coexistence plays the role of the thermodynamic field H in wetting transitions. Thus, one would assume that surface-induced disorder (as observed in Cu_3Au, Pb and Al) corresponds to complete wetting, since $H \neq 0$. It has been shown, however, that surface-induced disorder behaves like critical wetting ($H = 0$), as long as short ranged interactions dominate (Kroll and Lipowsky [1983]; see also the review by Dietrich [1988]). For a detailed discussion of this relationship we refer the reader to the contribution by Lipowsky in a following volume of this series.

b) All experiments on surface melting phenomena have been performed under UHV conditions, thus more or less away from equilibrium depending upon the saturation vapour pressure at the melting temperature. This undersaturation leads to permanent evaporation processes at the crystal-vacuum interface during the experiment which in turn should add to the equilibrium surface disorder. This is confirmed by a theoretical study within a Landau theory [Löwen and Lipowsky, 1991]: The amplitude ξ_d of the logarithmic growth (5.36) becomes enhanced under the presence of surface evaporation, $\xi_d^+ = \xi_d(1 + C v_{eva})$, where v_{eva} is the evaporation velocity, which can be estimated from the monolayer time (discussed in Sect. 2.4.3; see also A.11), and C is a constant. The effect is generally negligibly small.

As pointed out above, a simple surface melting rule based on free energy considerations is

$$\gamma_{sv} = \gamma_{sl} + \gamma_{lv} \tag{5.38}$$

("*Antonow's rule*") which states that the surface energy stays the same after the intercalation of a liquid film. While the liquid-vapour interfacial energy γ_{lv} is more or less precisely known from conventional surface tension measurements, the solid-liquid and solid-vapour interfacial energies are much less accessible to experiments, consequently, reliable values are very scarce. Experimental data for γ_{sl} mainly stem from grain-boundary grooving and homogeneous nucleation (for a review see [Woodruff, 1973]), whereby results obtained by the latter method suggest the socalled "*Turnball relation*" $\gamma_{sl} \simeq 0.45 L_m$ (L_m molecular latent heat of melting). Hilliard and Cahn [1958] already remarked that the low value of γ_{sl} is an indicator for the presence of a rather diffuse solid-liquid interface, a feature which will be addressed below in more detail (Sect. 5.2.3). The biggest uncertainty in the surface melting rule (5.38) is due to the circumstance that it is difficult to get an experimental hand on the solid-vapour interfacial energy γ_{sv}. Miedema [1978] showed that the difference between γ_{lv} and γ_{sv} is very small, as a rule of thumb, $\gamma_{sv} = 1.13 \gamma_{lv}$ for pure metals. Recently Kendall [1990] proposed to measure γ_{sv} electrically, however, this method only works in the case of conducting powder. A summary of the various experimental techniques to measure γ_{sv} and γ_{lv} as well as a table of γ_{sv} of some elements is given by Kumikov and Khokonov [1983]. Further useful references are found in the article by Shaler [1953]. Note here that γ_{sv} depends sensitively on the crystallographic orientation of the solid surface.

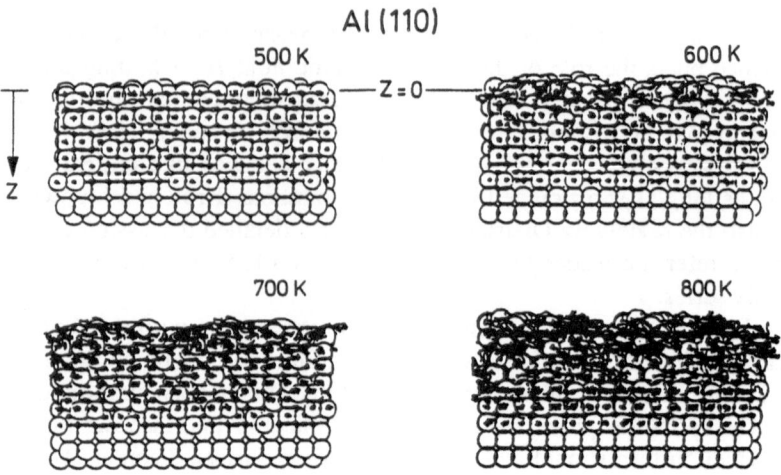

Al (110)

500 K   —Z = 0—   600 K

700 K   800 K

Fig. 5.15. Molecular dynamics calculation of the surface disordering at the Al (110) surface [Stoltze et al., 1989]

A pioneering molecular dynamics study of the surface melting phenomenon has been undertaken by Stoltze et al. [1989] at the Al (110) surface. Since realistic selfconsistent Al-potentials (which include to some extent many-body interactions) are used in this work, the obtained results can be considered as very reliable. Inspection of Fig. 5.15 which shows the calculated trajectories of the near-surface Al atoms at four different temperatures already tells us that extensive surface disorder emerges between 700 K and 800 K. Indeed, the deduced layer-LRO indicates that, at $T \simeq 735$ K, the top three layers are quasiliquid and that the thickness of this film grows upon further increase of the temperature.

5.2.2 Near-Surface Crystallinity

A large body of excellent experimental work on surface melting has been done by ion shadowing. The quantity measured with this technique is s *local disorder* around one atom. Another important information, however, for the understanding of the surface melting process is the crystalline *long range order* which is present in the visible disordered surface layers. This crucial information on the socalled "rest crystallinity" buried in the quasiliquid can in principle be provided by evanescent x-ray and neutron scattering.

Consider a liquid adjacent to a solid surface. The periodic substrate potential will imprint a modulation into the liquid with the Fourier components $G_\parallel$ of the solid. Intuitively it is clear that the associated inverse decay length $K(G_\parallel)$ in the modulated liquid should somehow scale with $|G_\parallel|$ (*"Higher Fourier components decay faster."*). $K^{-1}(G_\parallel) \equiv a(G_\parallel)$ has recently been discussed by Lipowsky et al.[1989] within a Landau theory and by Löwen et al.[1989] and Löwen and Beier [1990] within a density functional (DF) theory (see also Chernov and Mikheev [1988] who consider an effective interface model which includes the density

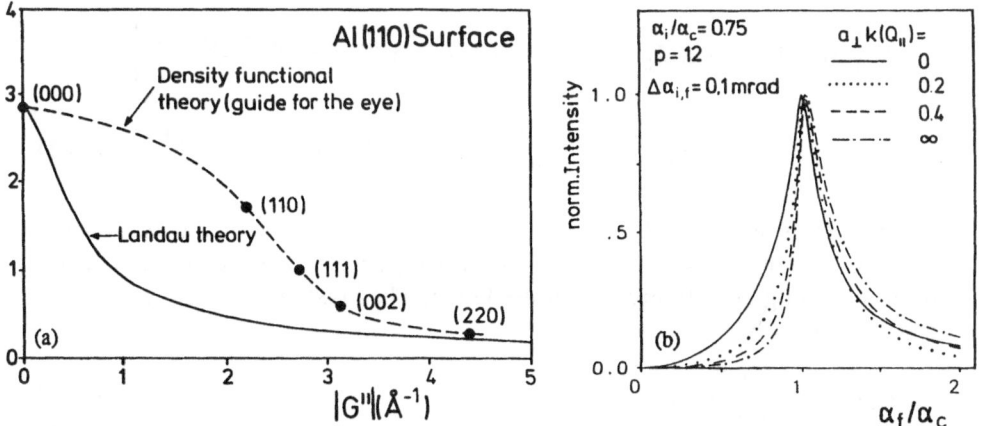

Fig. 5.16a,b. Rest crystallinity in the undercooled quasiliquid surface layer at the Al (110) surface. (a) Decay length $a(G_\parallel)$ of the solid Fourier components $G_\parallel$ from MFA (solid line; Lipowsky et al. [1989]) and from density functional theory [Löwen et al., 1989] (b) Normalized Bragg profiles for various values of $\kappa(G_\parallel) \equiv a^{-1}(G_\parallel)$ calculated for $\alpha_i/\alpha_c = 0.75$ and for a thickness $p = 12$ ML of the quasiliquid surface layer. Note that $\kappa = 0$ corresponds to a fully crystallized "liquid" (no surface melting) and $\kappa = \infty$ to the case of a thin bulk liquid. (All intensities are normalized to 1)

packing normal to the wall). The Landau theory yields the simple algebraic expression (full line in Fig. 5.16a)

$$a(G_\parallel) = \frac{a(0)}{\left[1 + \left(a(0)G_\parallel\right)^2\right]^{1/2}} \tag{5.39}$$

which, however, understimates $a(G_\parallel)/a(0)$ quite significantly for small values of $|G_\parallel| \neq 0$. The bulk correlation length $a(0)$ within the liquid, which governs the smectic density profile in the liquid adjacent to the solid, can be deduced from the liquid structure factor and is e.g. $a_{Pb}(0) = 4.7\,\text{Å}$, $a_{Ag}(0) = 2.75\,\text{Å}$ [Löwen and Beier, 1990]. The crystallinity profile $\varrho(Q_\parallel, z)$ at $T \leq T_m$ is

$$\varrho(Q_\parallel, z) = \begin{cases} \varrho_B(Q_\parallel) & \text{for } z \geq L \\ \varrho_B(Q_\parallel)\,e^{(z-L)\,K(Q_\parallel)} & \text{for } 0 \leq z \leq L \end{cases} \tag{5.40}$$

with $\varrho_B(Q_\parallel)$ as the lateral Fourier transform of the solid density-density correlation function and L being the thickness of the quasiliquid film. Experimental evidence for a remaining crystallinity in the surface melt has presumably been found in a LEED experiment from the Pb (110) surface [Prince et al., 1988]. However, it was also suggested by Stoltze [1989] that the observed anisotropy in the temperature dependence of (002) and (1$\bar{1}$0) Bragg reflections could also be due to a strong anharmonic Debye–Waller factor.

In a grazing angle x-ray or neutron scattering experiment from such a surface one obtains the square of the Bragg scattering amplitude

$$F_B\left(Q_\parallel, Q_z'\right) = \varrho_B(Q_\parallel) \left\{ \sum_{n=p}^{\infty} e^{iQ_z' a_\perp n} + \sum_{n=0}^{p-1} e^{(n-p)a_\perp K(Q_\parallel)} e^{iQ_z' a_\perp n} \right\}$$

$$= \varrho_B(Q_\parallel) \left\{ \frac{e^{iQ_z' pa_\perp}}{1 - e^{iQ_z' a_\perp}} + C_R\left(iR_z' a_\perp\right) e^{-pa_\perp K(Q_\parallel)} \right\} . \quad (5.41)$$

The first term in the bracket $\{\ldots\}$ accounts for the presence of a liquid surface layer of thickness $L = pa_\perp$ (see Dosch [1987]), while the second term arises from remaining cristallinity within the liquid. The *"rest crystallinity amplitude"* $C_R(x)$ is given by

$$C_R(x) = \frac{1 - e^{px}}{1 - e^{x}} . \quad (5.42)$$

In (5.41) $R_z' \equiv Q_z' - iK(Q_\parallel)$ is a new complex space frequency which contains the substrate modulation. Figure 5.16b shows the effect of the substrate modulation on the evanescent Bragg scattering for a fixed liquid film thickness $L = 12$ ML and various values of K: The sensitivity of $F_B\left(Q_\parallel, Q_z'\right)$ upon K in this Q_z'-range is apparently not very strong (note that $K = \infty$ corresponds to a liquid layer without any substrate modulation and $K = 0$ to a completely crystallized liquid).

5.2.3 Surface Sensitive X-Ray Scattering from Al and Pb

On the experimental side surface melting has been investigated by many different surface techniques such as optical emission [Stock, 1980], light scattering (see Bilgram [1987]), ion scattering (see van der Veen et al. [1988]), He scattering [Frenken et al., 1988], LEED [Prince et al., 1988; von Blanckenhagen et al., 1987], x-ray reflectivity [Gay et al., 1989] and evanescent x-ray scattering [Dosch et al., 1991b]. Experiments on surface melting have to be performed under UHV conditions in order to assure that an atomically clean surface is preserved. Therefore the solids Al and Pb are most favorable, because they have a reasonably low melting point ($T_{Al}^m = 933$ K, $T_{Pb}^m = 601$ K) and low vapour pressures around $p_m = 10^{-9}$ Torr. However, in grazing angle scattering experiments, substantial difficulties are faced in the mechanical preparation of the crystal surfaces: Owing to the softness of both materials the crystal structure at the surface suffers severe damage during polishing and the surface becomes usually very rounded at the edges. The lascivious oxidation of a Pb surface in air is a further problem. These facts thwarted for years several attempts in various research groups to perform thrustworthy grazing angle scattering experiments. In the following I will describe two surface sensitive x-ray scattering experiments on surface melting which have been performed by now, one on Pb (110) at the Daresbury synchrotron laboratory by Gay et al. [1989], the other on Al (110) at HASYLAB (Hamburg) by Dosch et al. [1991b].

The Al (110) surface preparation has been discussed in some detail in Sect. 2.4.4. It should be noted here that the relatively large roughness value of $\varrho = 10 \pm 2$ Å indicates that the Al (110) is probably already rough at room temperature. The in-plane mosaicity has been determined to be $\Delta\omega_\| = 0.1°$. The scattering experiments have been performed with a relatively large x-ray wavelength of $\lambda = 1.72$ Å [11] in order to counter the low Z of Al and to achieve a reasonable critical angle of $\alpha_c = 4.54$ mrad. In the temperature range between 500 K and 861 K the (002) α_f-Bragg profiles have been recorded at $\alpha_i/\alpha_c = 0.81$ by integrating the in-plane Bragg law via transverse scans.

The data and the final fits are summarized in Fig. 5.17 and reveal a distinct temperature-dependent decay of the evanescent Bragg intensity which is ascribed to two (not necessarily independent) mechanisms:

a) *A near surface thermal Debye–Waller factor* which is z-dependent close to the surface,

$$2B(z, T)/Q_\|^2 = \langle u_\|^2(z, T)\rangle \quad , \tag{5.43}$$

and includes surface anharmonicities and also to some extent surface roughening effects. Notice that anharmonic lattice vibrations of the surface atoms at the Al (110) surface have been detected in molecular dynamics simulations [Stoltze et al., 1988], they exhibit in a leading order a quadratic temperature dependence [Maradudin and Flinn, 1963].

b) *The appearance of a quasiliquid surface layer* which does not contribute to the evanescent Bragg intensity (see also Sect. 2.2.3). The effective thickness is described by $L(T) = p(T)a_\perp$ with the integer number $p(T)$ as a free fitting parameter to the α_f-profiles.

The evanescent Bragg scattering amplitude then reads

$$F_B(Q_z', T) \propto \sum_{n=p(T)}^{\infty} e^{-B(a_\perp n)} e^{-iQ_z' a_\perp n}$$

$$\simeq e^{-\langle B\rangle_\Lambda} e^{iQ_z' p(T)a_\perp} \frac{1}{1 - e^{iQ_z' a_\perp}} \quad , \tag{5.44}$$

where $\langle 2B\rangle_\Lambda \equiv \langle u_\|^2(T)\rangle_\Lambda Q_\|^2$ is the average lateral Debye–Waller factor within the illuminated surface layer of $\Lambda_{\text{eff}} \simeq 70$ Å (see (2.39)). Thus, by analyzing the α_f-profiles of the evanescent Bragg scattering, as provided by the experimental scheme described above, the two temperature-dependent effects, the average near-surface Debye–Waller factor $\langle u_\|^2(T)\rangle_\Lambda$ and the appearance of a quasiliquid surface layer of microscopic thickness $p(T)a_\perp$ can be obtained. A trivial (but generally small) temperature effect is the shift of the critical angle upon thermal expansion of the crystal. From (2.6) it follows that

$$a_c^{-1} \, d\alpha_c/dT = -\tfrac{3}{2} a_0^{-1} \, da_0/dT \simeq -\tfrac{3}{2} \gamma_T \tag{5.45}$$

(γ_T = linear thermal expansion coefficient), i.e. for Al with $\gamma_T = 2.2 \times 10^{-5}$ K^{-1}

---

[11] A much longer wavelength would entail a too large air absorption.

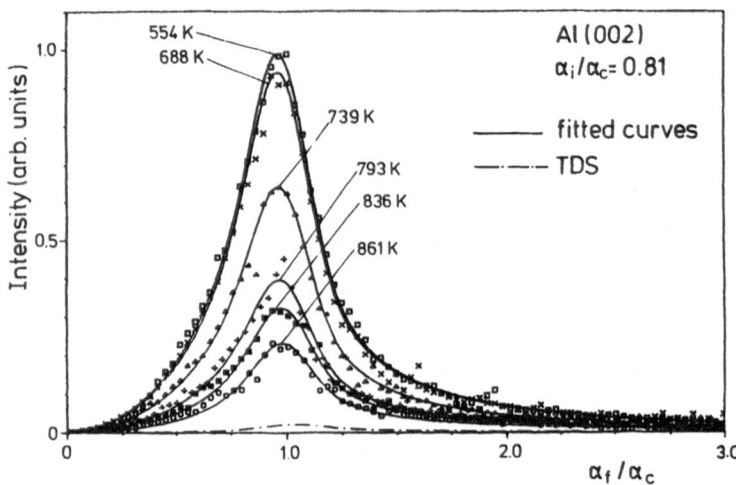

Fig. 5.17. Temperature dependence of the α_f-profiles of the in-plane-integrated (002) Bragg scattering. The full lines are the best fits within the DWBA using $\langle u_\|^2(T)\rangle$ and $p(T)$ as independent fit parameters (see text)

[Landolt–Boernstein, 1971] the critical angle decreases by roughly 2 % when the sample is heated from room temperature to its melting temperature (T_m = 933.5 K).

In the scattering experiment the α_f-profiles of the $\Delta\omega_\|$-integrated (002) Bragg intensity were measured at $\alpha_i/\alpha_c = 0.81$ for various temperatures between 518 K and 861 K (Fig. 5.17). Based on (5.44) the changes in the *Bragg intensity* and *Bragg profile* have been analyzed and gave the full lines as the best fits with $p(T)$ and $\langle u_\|^2(T)\rangle_\Lambda$ as free fitting parameters. The dashed curve shows the calculated α_f-profile of the temperature diffuse scattering contribution at $T = 861$ K. This scattering has been included into the theoretical model at each temperature assuming a standard high temperature approximation. The resulting values for $p(T)$ and $\langle u_\|^2(T)\rangle_\Lambda$ are shown in Fig. 5.18 versus sample temperature indicating three different regimes:

a) 520 K $< T <$ 690 K: The temperature dependence of the (002) Bragg intensity and the associated α_f-profiles can completely be explained using a harmonic Debye–Waller factor (dashed line in Fig. 5.18), $\langle u_\|^2(T)\rangle_\Lambda = \alpha_\| \cdot T$ and $\alpha_\| = 4.8 \times 10^{-5}\,\text{Å}^2/\text{K}$. The bulk Debye temperature $\theta = 394$ K implies $\alpha_B = 3.5 \times 10^{-5}\,\text{Å}^2/\text{K}$, thus, $\alpha_\|$ appears distinctly enhanced compared to the bulk in apparent agreement with the conclusions drawn from the LEIS measurements [Denier v.d.Gon et al., 1990].

b) 690 K $< T <$ 740 K: In this temperature regime a pronounced change in the temperature dependence of the evanescent x-ray intensity is experienced very similar to the LEED observation. The least-squares fit analysis suggests a distinct deviation of $\langle u_\|^2(T)\rangle_\Lambda$ from harmonic behaviour at $T \simeq 690$ K.

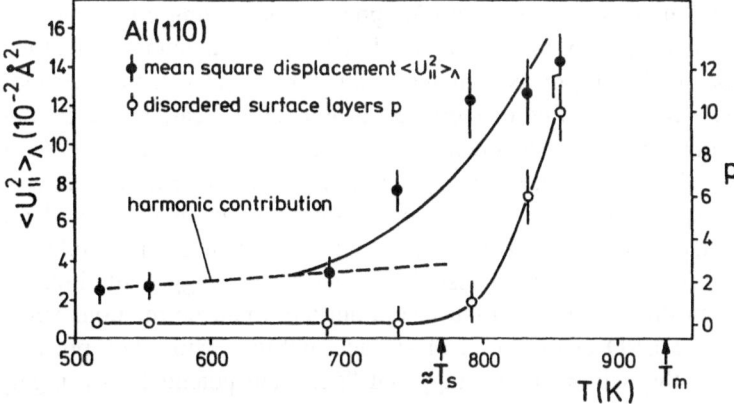

Fig. 5.18. Temperature dependence of the fit parameters $\langle u_{\parallel}^2(T) \rangle$ and $p(T)$. T_s denotes the onset of surface melting deduced from the condition $p(T) = 1$. The full line indicates the regime of harmonic lattice vibrations of the atoms near the (110) surface

c) $T > 770$ K: A strong decay of the evanescent Bragg intensity is found for $\alpha_f/\alpha_c < 1$ which can best be reproduced in the fit by introducing surface layers which exhibit no noticeable Bragg scattering any more. The effective number $p(T)$ of these "dead layers" increases steadily with temperature as shown in Fig. 5.18.

According to these results the following scenario for emerging surface disorder is proposed: At $T \simeq 690$K a first precursor of the starting surface disorder occurs at the onset of anharmonicities detectable in $\langle u_{\parallel}^2(T) \rangle_A$. One has to conclude from this observation that prior to the actual surface melting a strong static and/or dynamic disorder of the surface takes place which finally leads to the emergence of quasiliquid surface layers. This confirms the conclusions drawn from molecular dynamics calculations that strong displacements of the surface atoms play a preparatory role in the surface melting process. The onset temperature for surface melting, as characterized by $p = 1$, is estimated from this study to be $T_s \simeq 770$K$\pm$ 25 K, while from ion shadowing measurements one finds $T_s = 815$ K. Within this x-ray scattering study the transient temperature regime between the onset of surface anharmonicities and the onset of surface melting is considerably blurred out. This is mainly owing to the sensitivity of the near-surface LRO to all kinds of disordering mechanisms, as surface roughness, anharmonic lattice vibrations and surface melting itself. The associated evanescent Debye–Waller factors which account for the different types of surface disorder are different functions of $Q_{\parallel}$, Q'_z, etc. and can thus be distinguished. However, it turns out that one depends on a detailed χ^2-analysis of the data for a proper disentanglement. This complex situation also entails the comparatively large error bars in the fitting parameters shown in Fig. 5.18 and in the estimate of T_s. At $T = 860$K the *effective thickness* of the surface layer which contains practically no more Fourier components of the solid has increased to about $9 - 10$ atomic layers. Since the contributions from surface anharmonicities and eventually present roughening effects are not subtracted, this thickness has to be compared with the *total* number of atoms visible in the ion shadowing measurements, $p(860\,\mathrm{K}) = 7$ [Denier von der Gon et al., 1990]. The origin of this discrepancy is not quite clear, however, one has to keep in mind that ion shadowing experiments monitor the increase of *local disorder*, while the evanescent Bragg scattering is sensitive to the decrease of *long range order*. Since short range correlations remain in the disordered surface layers when LRO has already disappeared, it seems not unplausible that ions should measure a smaller thickness than evanescent Bragg scattering. We can conclude from this that the near-surface atom-atom pair correlations suffer a significant loss of LRO at the onset of surface melting, while they preserve – probably due to the action of the staggered "substrate potential" – a highly correlated local order.

I want to end the discussion of the Bragg scattering with a note on the rest crystallinity which, in principle, could be obtained from the α_f-profiles of the Bragg intensity. The DF theory predicts $K(002) \simeq 1.5\,\mathrm{\mathring{A}}^{-1}$ for Al (110) (Fig. 5.16a). This inverse decay length, however, is already too large to give a striking deviation from $K = \infty$ (see Fig. 5.16b). The task to extract such details of the OP profile from α_f-profiles of evanescent Bragg scattering is certainly one of the future experimental challenges. Such attempts, however, can only be promising, when the surface quality of these crystals, i.e. the surface waviness is further improved.

Probably the most important piece of information needed for a microscopic understanding of the surface melting process is buried in the (still unknown) structure of the undercooled quasiliquid surface layer. Surface melting theories are in the stage to predict the normal density profile $\varrho(Q_{\parallel}, z)$ within the disordered surface layer, in particular the DF theories which use the liquid structure factor for a reliable guess of the many body interactions which are important (see Stoltze et al. [1989]). So far no thrustworthy experimental determination of the lateral structure factor of the quasiliquid has been reported, on the theoretical side one promising approach to attack this task has been presented by Reiter and Moss [1986] who calculated the x-ray structure factor of a liquid monolayer intercalated between staggered substrate potentials within a nonlinear response theory. It would be highly desirable to apply/extent this theory to the surface melting phenomenon.

Asymptotically, for $T = T_m$, the surface melt turns into a thermodynamically stable bulk liquid which is characterized by the disappearance of all solid density Fourier components and the appearance of a new average density. Thus, one expects that upon approaching T_m an average density profile across the sample-vapour interface should develop with slowly appearing signatures of a liquid layer. Experimentally this is best observed by specular reflection of x-ray or neutrons (see e.g. Lekner [1991] and Fig. 5.19a). A first x-ray experiment of this kind has been carried out (at the synchrotron radiation laboratory SRS, Daresbury) on the Pb (110) surface [Gay et al., 1989]. Figure 5.19b shows the observed normalized reflectivity $I(Q_z)/I_F(Q_z)$[12] at $T = T_m - 0.2$ K together with a fit assuming a quasiliquid surface layer with the layer thickness L and the average density ϱ_{QL} of the quasiliquid as free fitting parameters (full line). One gets $L \simeq 20$ Å which agrees very well with the ion shadowing experiments and a density $\varrho_{QL} = (1 - 0.039)\varrho_s$ (ϱ_s is the solid density at T_m) which is close to the density of the bulk liquid $(\varrho_s - \varrho_l)/\varrho_s = 3 \times 10^{-2}$ (see Borelius [1958]), and in fact appears slightly less dense than the bulk liquid. This density difference $(\varrho_s - \varrho_l)$ enters – as discussed above – the long ranged interactions (see 5.35). Since $(\varrho_s - \varrho_l) > 0$, the associated contribution $C\varrho_l(\varrho_s - \varrho_l)/L^{-2}$ (5.35) is repulsive, i.e., favours surface melting, one expects a crossover from the soft $\ln t$-divergence of the quasiliquid layer to a $t^{-1/3}$-divergence [Lipowsky, 1987; Trayanov and Tossati, 1988], when the film has reached a certain thickness L. This crossover has indeed been observed at the Pb (110) surface at $T = T_m - 0.3$ K [Pluis et al., 1989]. At this point I want to come back to the *"superlattice surface melting"* at the Cu$_3$Au (001) surface. As discussed above the associated density difference is found to be $(\varrho_{ord} - \varrho_{dis}) \simeq 3 \times 10^{-3}$, thus, about 10-times smaller than $(\varrho_s - \varrho_l)$ of Pb (see Sect. 5.1.1). Therefore, one would expect this $\ln t - t^{-1/3}$-crossover to happen in Cu$_3$Au only very shortly before the transition temperature is reached.

From the reflectivity profile (Fig. 5.19b) Gay and co-workers [1989] also tried to deduce the interfacial widths between the vapour and the quasiliquid ($\xi_{\perp lv}$) and between the quasiliquid and the solid ($\xi_{\perp sl}$) from the reflectivity data and find $\xi_{\perp lv} \simeq 3$ Å and $\xi_{\perp sl} \simeq 9$ Å. Apparently the solid-quasiliquid interface is rather diffuse, a conclusion which is confirmed by the x-ray experiment on Al (110) and by a lattice theory of the surface melting of Lennard–Jones crystals [Trayanov and Tosatti, 1988]. The relatively small value of $\xi_{\perp lv}$ seems to be a typical interfacial width for a liquid-metal/vapour interface [Allen and Rice, 1977].

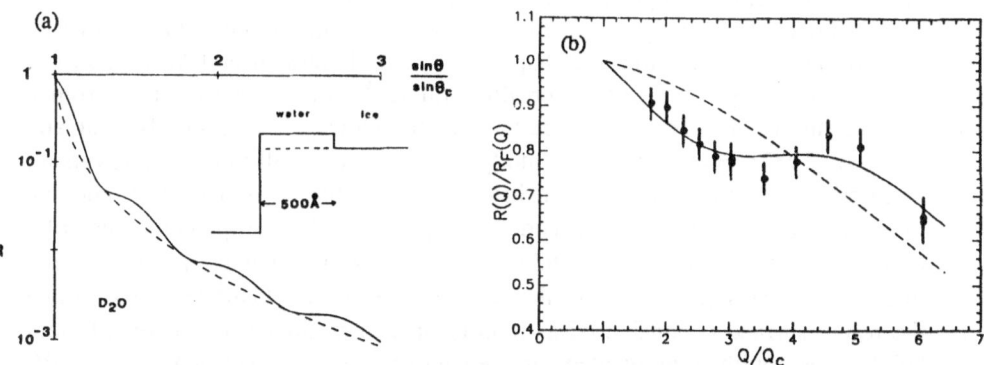

Fig. 5.19a,b. (a) Calculated neutron reflectivity due to a 500 Å thick layer of water on ice (D$_2$O) [Lekner, 1991]; the dashed curve is the reflectivity due to ice alone, the inset shows the associated scattering length density profile (not on scale); (b) observed normalized x-ray reflectivity profile from Pb (110) at $T = T_m - 0.2$ K [Gay et al., 1989]

---

[12] $I_F(Q_z)$ is the Fresnel reflectivity of the ideal surface.

For the near future there remain many open questions which can be answered by grazing angle scattering, such as:

– Which planar structure has the undercooled quasiliquid surface layer?
– How does the rest crystallinity decay within this wetting layer?
– Is the disordered surface layer liquid- or glass-like?

In particular the last question may be seen in view of the *Kauzmann–criterion* [Kauzmann, 1948] according to which an undercooled liquid whose entropy is less than that of the crystalline phase must transform into a glass state ("*entropy catastrophy*"). Applied to surface melting this would mean that the undercooled quasiliquid layer is first glass-like at T_s when $T_s < T_g$. In Al by the way the *Kauzmann–criterion* would yield a glass transition temperature of $T_g = 225$ K [Fecht and Johnson, 1988] which would be well below the observed onset temperature for surface melting, consequently a substrate-modulated undercooled surface-liquid would be expected to occur. He-scattering measurements of the surface diffusion of Pb atoms at the Pb (110) surface indicate a liquid-like diffusion coefficient of the disordered surface atoms at the onset temperature for surface melting [Frenken et al., 1988].

5.3 Surface Roughening

The loss of the lateral positional LRO as it occurs during surface melting may be accompagnied by the precursory destruction of the height-height correlations of the surface atoms ("*surface roughening*"). In fact most of the crystal surfaces are rough during melting with only a few exceptions, where facets have been observed during the melting process. Thus, any discussion of surface melting should take surface roughening effects into account.

The concept of surface roughening has been introduced in the study of crystal growth and of the equilibrium shape of crystals [Rottman and Wortis, 1984]. The classical experiment has been done on He^4, where the transition from a smooth surface to a rough surface has unambiguously been seen [Balibar and Castaign, 1985]. However, its thermodynamic nature is still rather controversial: Theoretically various types of transitions are thinkable from critical roughening to an infinite order (Kosterlitz–Thouless type) transition. Apparently the order of the transition depends delicately upon the model and upon the underlying interactions (see Rys [1986]). In the widely used discrete Gaussian (DG) model the rough surface is considered as closely packed parallel columns of different heights $h(r_\parallel)$ with an associated surface energy $\sim J_R/2 \sum_j [h(r_{\parallel j}) - h(r_{\parallel j+1})]^2$. It turns out that the DG model is isomorphic to the $2d$ Coulomb gas [Chui and Weeks, 1976], thus belongs to the $2d$ Kosterlitz–Thouless universality class. As a further complication there is the effect of surface reconstruction on the roughening behaviour: One knows that the clean (110) surface of light metals, like Cu and Ag, does not reconstruct, while the clean (110) surface of heavier metals, like Au and Pt, exhibits a *"missing-row"* reconstruction. Trayanov et al. [1989] argued that the former surfaces undergo a socalled *"missing-row"*-roughening transition (see also Van Beijeren [1977]). At any rate there exist reliable theories which predict a smooth surface for low temperatures and a rough surface for high temperatures, thus, there seems to be no doubt about the existence of a roughening transition (see Villain et al.[1985]), at least for $d = 3$[13]. At temperatures $T < T_R$ (T_R is the roughening transition temperature) the height-height correlation function $g_\perp(r_\parallel)$ remains bounded, while for $T > T_R$, resulting from the copious proliferation of atomic steps at the surface, $g_\perp(r_\parallel)$ diverges (see Weeks [1980] and Fig. 5.20),

$$\lim_{r_\parallel \to \infty} g_\perp(r_\parallel) \propto \begin{cases} C(T) & \text{for } T < T_R \\ R(T) \ln |r_\parallel| & \text{for } T \geq T_R \end{cases}, \qquad (5.46)$$

where $C(T) = \xi_\perp^2(T)$ and $R(T)$ are temperature dependent amplitude [Villain et al., 1985]. By glancing angle x-ray and neutron scattering the roughening can be observed in a direct way: The associated (diffuse) surface scattering signal is given by $S_{\text{dif}}^\varrho(Q')$ (3.43) which has been derived for neutron scattering, but holds as well for x-rays (after replacing the neutron-matter interaction by the appropriate x-ray-matter coupling constant). In the expression (3.43) we removed any eventually present Bragg scattering by subtracting "1", the total surface scattering is thus

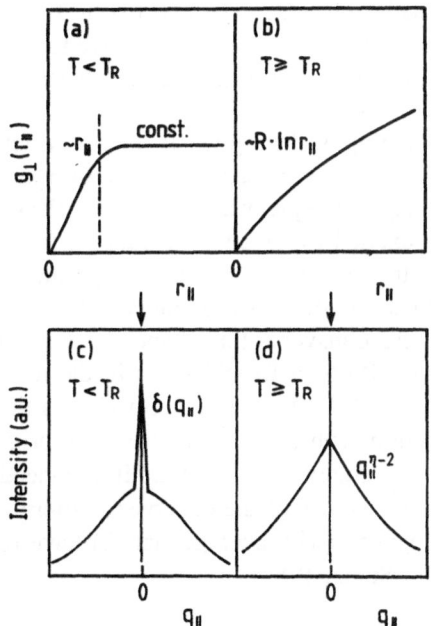

Fig. 5.20a-d. Height-height correlation functions $g_\perp(r_\parallel)$ and surface scattering signal: (a) $g_\perp(r_\parallel)$ for $T < T_R$: $g_\perp(r_\parallel)$ is constant for large $r_\parallel$, while the linear increase for small $r_\parallel$ is associated with local steps; (b) $g_\perp(r_\parallel)$ for $T \geq T_R$: $g_\perp(r_\parallel)$ exhibits a logarithmic divergence; (c),(d) surface scattering intensities $I(Q_\parallel)$ for case (a) and (b), respectively

---

[13] Within the DG model it appears quite plausible that the surface of low dimensional systems ($d < 3$) is always rough ($T_R = 0$), while the surface of high dimensional systems ($d > 3$) never roughens ($T_R = T_c$).

$$S(\boldsymbol{Q}_\parallel, q_z') \propto \iint d^2 r_\parallel \, e^{-|q_z'|^2 g_\perp(r_\parallel)} e^{i Q_\parallel r_\parallel} \quad . \tag{5.47}$$

Inserting (5.46) into (5.47) one finds in a straightforward way that $S(\boldsymbol{Q}_\parallel, q_z')$ is composed out of a δ-function and a Lorentzian in $\boldsymbol{Q}_\parallel$ for $T < T_R$ (Fig. 5.20c) and takes on a powerlaw form for $T \geq T_R$ (see Fig. 5.20d),

$$S(\boldsymbol{Q}_\parallel, q_z') \propto |Q_\parallel|^{\eta(T, q_z') - 2} \quad . \tag{5.48}$$

The roughness exponent

$$\eta(T, q_z') \equiv R(T) \, q_z'^2 \tag{5.49}$$

depends on the reduced momentum transfer $q_z' = G_{hkl} - \kappa_z'$. According to Villain et al. [1985] $R(T_R) = (a_\perp/\pi)^2$, thus, the exponent $\eta(T_R, q_z')$ at the roughening transition variies from 0 at $q_z' = 0$ (at Bragg points) to 1 at $q_z' = \pi/a_\perp$ (at zone boundaries)[14].

As already mentioned in Sect. 3.4 the truncation of the translational invariance at the surface shows up in a scattering experiment as socalled *"asymptotic Bragg wings"* (or *"crystal truncation rods"* [Robinson, 1986]) normal to the surface (z-direction) which exhibit a q_z^{-2}-powerlaw decay for a mathematical surface [Andrews and Cowley, 1985]. Of particular interest are the afore-mentioned Brillouin zone boundary points along the z-direction ("X" in Fig. 5.21a), where the Bragg scattering amplitudes between two alternate planes are exactly out of phase, i.e. waves generated in the bulk cancel each other. The asymptotic Bragg wings at these "X-points" are hence most surface sensitive and responsive to tiny changes in the differential scattering amplitude between neighbouring planes, as created by surface roughness or surface segregation.

Held et al. [1987] investigated the roughening of the Ag (110) surface which belongs (as Cu (110) discussed below) to the class of surfaces which display no reconstruction. The applied experimental technique was asymptotic Bragg scattering at the ($1\bar{1}1$)-point which can be excited by grazing incidence diffraction (Fig. 5.21a). The lateral line shape of the asymptotic intensity is predicted to turn from a δ-function associated with a smooth surface into a powerlaw scattering (5.48) for $T \geq T_R$. The radial and transverse line shapes have been measured with synchrotron radiation (at the NSLS in Brookhaven) in the temperature range between 373 K and 823 K (Fig. 5.21b). At low temperatures the observed $S(\boldsymbol{Q}_\parallel, q_z')$ is essentially a δ-function with small powerlaw tails originating from local steps, while at high temperatures $S(\boldsymbol{Q}_\parallel, q_z')$ exhibits indeed only powerlaw scattering. In the quantitative analysis of the observed scattering profiles finite-size effects due to surface domains (with average size L) have to taken into account. This leads to [Dutta and Sinha, 1981]

---

[14] Note that the scattering law (5.48) is similar to the one present in two-dimensional crystals whose positional LRO is destroyed at $T > 0$ by long-wavelength phonons (see Imry and Gunther [1971]).

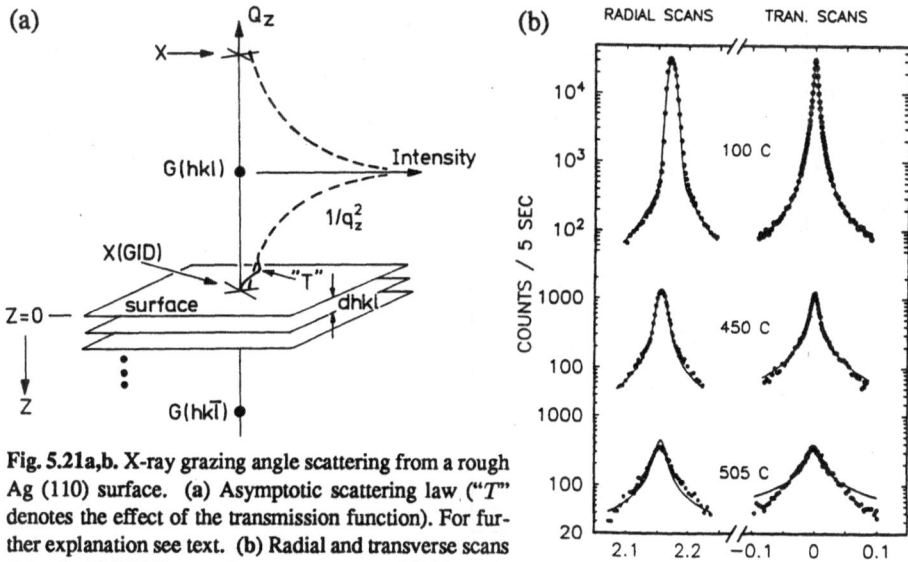

Fig. 5.21a,b. X-ray grazing angle scattering from a rough Ag (110) surface. (a) Asymptotic scattering law ("T" denotes the effect of the transmission function). For further explanation see text. (b) Radial and transverse scans through X [GID] at various temperatures [Held et al., 1987]

$$S(Q_\parallel, q_z') \propto (L/\xi)^{2-\eta(T, q_z')} \begin{cases} \phi(u; 1; -w) & \text{for a Gaussian} \\ F(u; u; 1; -w) & \text{for a Lorentzian} \end{cases} \quad (5.50)$$

with $u \equiv 1 - \eta/2$, $w \equiv q_\parallel^2 L^2/4\pi$ and η given by (5.49). In (5.50) $\phi(\alpha; \gamma; z)$ is the degenerate hypergeometric function ("*Kummer function*") and $F(\alpha; \beta; \gamma; z)$ the hypergeometric function. Since for $\eta = 0$, the scattering law $S(Q_\parallel, q_z')$ converges to a pure Gaussian and Lorentzian, respectively (see B.9 and B.10), the course of the roughening can very conveniently be modelled by the exponent η (full lines in Fig. 5.21b), by this the roughening transition of this surface has been localized at $T_R(110) = 723 \pm 25$ K which is significantly below the melting temperature $T_m = 1234$ K (i.e. $T_R(110)/T_m = 0.59$). A crucial test of the scattering law (5.48) to describe the roughening of a surface is the experimental study of the distinct q_z'-dependence of $\eta(T, q_z')$ (5.49). Such an x-ray scattering study on Ag (110) has been reported by Robinson et al. [1991], who, however, find that $S(Q_\parallel, q_z')$ deviates increasingly more from the predicted powerlaw behaviour (5.48) the more q_z' deviates from the zone boundary point: At room temperature, below the roughening temperature (which was determined in this study to be $T_R(110) = 790 \pm 20$ K, thus $T_R/T_m = 0.65$), the data show a pronounced two-component lineshape for $q_z' \neq 0$ (Fig.5.22). This implies that the Ag (110) surface is segregated into flat, (110) oriented regions (giving rise to the sharp component aligned with the crystallographic (110) direction) and rough regions inclined by an angle $\alpha(300 \text{ K}) = 0.25°$ with respect to the (110) direction (giving rise to the broad component). The observed temperature dependence of the angle α provides a strong evidence that the roughening at this surface takes place by a gradual replacement of the (110) facetted regions by the rough phase. The

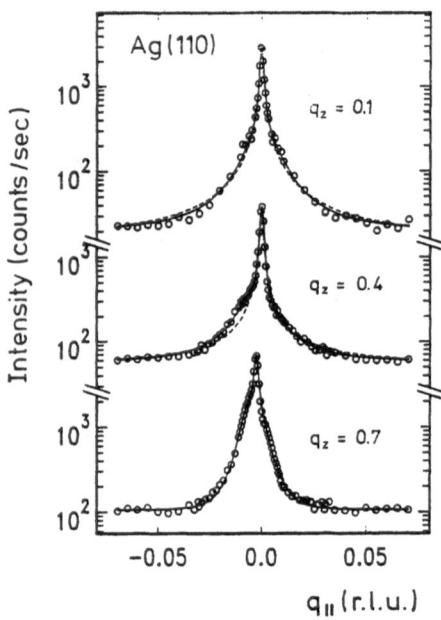

Fig. 5.22. Radial scans along $(1\bar{1}0)$ of the $(1, 1, q_z)$ crystal truncation rod of Ag(110) at room temperature for $q_z = 0.1$, 0.4 and 0.7 [Robinson et al., 1991]. The full line is the best fit using a two-component lineshape, the dashed assumes the powerlaw behaviour of (5.48)

difference between the two values of T_R (110) in the x-ray studies by Held et al. and Robinson et al. may be explained by the different miscuts $\Delta\phi$ of the (110) surfaces which were $\Delta\phi \leq 0.3°$ and $\Delta\phi \leq 0.15°$ in the former and latter experiment, respectively.

A further x-ray scattering experiment has been performed on the Cu (110) surface [Mochrie, 1987] yielding a roughening transition of $T_R(110) \simeq 873$ K ($T_m = 1356$ K). In this study, however, only a strong decrease of the asymptotic Bragg intensity has been recorded and no change in the lateral intensity profile. It was remarked by Zeppenfeld et al. [1989] that the analysis of He-scattering data from Cu (110) do not yield a roughening of this surface for temperatures $T \leq 900$ K and rather points to an anomalous increase of the mean-square displacement of the surface atoms for $T \geq 550$ K. This conclusion has been confirmed by a recent *I*mpact *C*ollision *I*on *S*cattering *S*pectroscopy (ICISS) study on the same same system [Dürr et al., 1991]. In addition one finds there indications of a structural surface transition at roughly 1000 K which is assumed to be either a surface roughening (i.e. at $T_R(110)/T_m \simeq 0.75$) and/or surface melting effect, a conclusion which in turn is very similar to the one drawn from the x-ray observation at the Al (110) surface [Dosch et al., 1991b], where a strong disordering of the surface apparently occurs at $T_s(110)/T_m \simeq 0.82$ interpreted as the onset of surface melting (see Sect. 5.2.3).

A quite different grazing angle scattering experiment has been reported by Liang et al. [1987] who investigated a stepped Cu (113) surface. At such "high Miller index" surfaces step rows are already present for $T < T_R$. At T_R the exuberant appearance of kinks leads to a dramatic meandering [Villain et al.,1985]. Since the energy cost to create a kink is noticeably lower than that to create a step, T_R associated with the (113) surface is expected to be lower than T_R measured at the (110) surface. This "hkl-dependence" of T_R can also be deduced from the RG roughness criterion (see Weeks [1980])

$$T_R(hkl) = \frac{2\varepsilon_S a_0^2}{\pi k_B(h^2 + k^2 + l^2)} \tag{5.51}$$

(with ε_S as the surface free energy and k_B as the Boltzmann constant) according to which T_R decreases with increasing Miller indexing of the surface (note also that within the "$d = 3$"-DG model $T_R = 1.2\, J_R/k_B$). At low temperatures the equilibrium surface is an ordered stepped surface which gives rise to grazing angle superlattice reflections, at and above the roughening transition the step superlattice disappears. Along the (113) surface the step-step distance is $d_s = 4.239$ Å, thus a step superlattice peak is expected at $G_s = 2\pi/d_s = 1.48$ Å^{-1}. Experimental results of the temperature dependence of the step superlattice G_s (obtained at the NSLS) are shown in Fig. 5.23. The temperature at which the step superlattice intensity disappears is identified with T_R yielding a roughening transition $T_R(113) = 620 \pm 10$ K, thus, $T_R(113)/T_m = 0.46$.

Robinson et al. [1990] recently investigated the roughening of the Ni (113) surface by measuring the temperature dependence of the asymptotic Bragg scattering at grazing incidence similar to the experiment by Held et al. [1987]. The in-surface lineshape $S(Q_\parallel, q_z')$ (5.47) of the observed scattering was analyzed and yields a roughening temperature of $T_R(113) = 770\pm30$K, i.e. $T_R(113)/T_m = 0.45$ which is the same ratio as before in the Cu (113) experiment. This is a first experimental evidence that $T_R(hkl)$ and T_m are universally related for metals with the same crystal and surface structure.

In molecular dynamics studies of Cu (110) and Al (110) surfaces Ditlevsen et al. [1991] found the thermal generation of adatom-vacancy pairs at the outermost surface layer which seems to occur prior to the onset of surface melting. This is another strong evidence that surface roughening (here in the form of the adatom population) and surface melting are correlated effects. The future challenge for the experimentalists is to perform *at one sample* more comprehensive studies

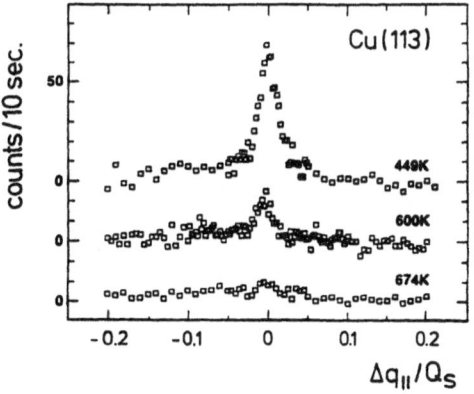

Fig. 5.23. Grazing angle "step-superlattice" scattering from the Cu (113) surface [Liang et al., 1987]

of the various surface scattering signals: The simultaneous investigation of the temperature dependence of reflectivity profiles, surface truncation rod scattering, surface disorder-induced diffuse scattering and evanescent Bragg scattering is quite an experimental effort, however, it will certainly provide further new insight into the fundamental role of the crystal border in the solid-liquid transition.

Appendix

A: Useful Relations

1. Conversion: Energy – Wavelength

X-rays: $\qquad \lambda\,[\text{Å}] = \dfrac{12.4}{E\,[\text{keV}]}$ $\hfill$ (A.1)

Neutrons: $\quad \lambda\,[\text{Å}] = \dfrac{9.05}{\sqrt{E\,[\text{meV}]}}$ $\hfill$ (A.2)

2. Critical Angle from Mass Density

X-rays: $\qquad \alpha_c\,[\text{mrad}] = 2.317\,\lambda\,[\text{Å}]\,\sqrt{\varrho\,[\text{g/cm}^3]\,Z/A}$ $\hfill$ (A.3)

Neutrons: $\quad \alpha_c\,[\text{mrad}] = 1.380\,\lambda\,[\text{Å}]\,\sqrt{\varrho\,[\text{g/cm}^3]\,b_{coh}\,[\text{fm}]\,/A}$ $\hfill$ (A.4)

(A is the mass number in the periodic table.)

3. Minimal Scattering Depth from Mass Density

X-rays: $\qquad \Lambda_{min}\,[\text{Å}] = \dfrac{34.3}{\sqrt{\varrho\,[\text{g/cm}^3]\,Z/A}}$ $\hfill$ (A.5)

Neutrons: $\quad \Lambda_{min}\,[\text{Å}] = \dfrac{57.6}{\sqrt{\varrho\,[\text{g/cm}^3]\,b_{coh}\,[\text{fm}]\,/A}}$ $\hfill$ (A.6)

4. Conversion: Mass Absorption Coefficient μ/ϱ — Absorption Cross Section σ_a:

$$\sigma_a\,[\text{bn}] = 1.6726\,A\,\frac{\mu}{\varrho}\,[\text{cm}^2/\text{g}] \hfill (A.7)$$

5. Al-Absorber for X-rays:

The required thickness d_{Al} of Al to attenuate a monochromatic x-ray intensity by a factor of f_{abs} is given by:

$$d_{Al} \, [\text{mm}] \simeq 0.611 \, \lambda \, [\text{Å}]^{-2.88} \, \log_{10} f_{abs} \tag{A.8}$$

(Note the rule of thumb $\mu \sim \lambda^{-3}$.)

6. Absorber for Thermal Neutrons:

In the absence of (n, γ) resonances the absorption cross section for thermal neutron follows the well-known $1/v$-law (v is the neutron velocity).

7. Wavelengths of X-Ray Absorption Edges:

The wavelengths associated with the K- and L-absorption edges of element Z are numerically (with an error less than 1%) given by

$$K\text{-edge:} \quad \lambda_K \, [\text{Å}] \simeq 1488(Z-1)^{-2.099} \quad \text{for } 13 \leq Z \leq 70 \tag{A.9}$$

$$L\text{-edge:} \quad \lambda_L \, [\text{Å}] \simeq 29935(Z-2)^{-2.365} \quad \text{for } 35 \leq Z \leq 100 \tag{A.10}$$

8. Monolayer Time [Vaughan, 1986]:

$$\tau_{ML} \, [\text{s}] = 3.2 \times 10^{-6}/p \, [\text{mbar}] \tag{A.11}$$

B: Table of Integrals and Functions

In the following I summarized some integrals and functions which occur in this review and may be helpful for the reader in the actual application of the presented results. They are found in the "Table of Integrals, Series and Products" by Gradshteyn and Ryzhik [1965].

$$\int_{-\infty}^{\infty} e^{-p^2 x^2 \pm qx} \, dx = \frac{\sqrt{\pi}}{p} \, \exp(q^2/4p^2) \quad (p > 0) \tag{B.1}$$

$$\int_0^{\infty} x J_0(xy) \, \frac{dx}{\sqrt{a^2 + x^2}} = e^{-ay}/y \quad (y > 0, \, \text{Re}\{a\} > 0) \, , \tag{B.2}$$

where $J_0(z)$ is the Bessel function of the first kind.

The complex Laplace transformation of order parameter profiles $m(x)$ usually involves integrals of the form:

$$\int_0^{\infty} x^{\nu-1} e^{-\mu x} \, dx = \mu^{-\nu} \Gamma(\nu) \quad (\text{Re}\{\mu\} > 0, \, \text{Re}\{\nu\} > 0) \tag{B.3}$$

$$\int_0^u x^{\nu-1} e^{-\mu x}\, dx = \mu^{-\nu}\, \gamma(\nu, \mu u) \quad (\text{Re}\{\nu\} > 0) \tag{B.4}$$

$$\int_u^\infty x^{\nu-1} e^{-\mu x}\, dx = \mu^{-\nu}\, \Gamma(\nu, \mu u) \quad (\text{Re}\{\mu\} > 0,\ \text{Re}\{\nu\} > 0) \tag{B.5}$$

with

$$\Gamma(\alpha) \equiv \int_0^\infty e^{-t} t^{\alpha-1}\, dt \quad (\text{``Gamma function''}),$$

$$\gamma(\alpha, x) \equiv \int_0^x e^{-t} t^{\alpha-1}\, dt \quad (\text{``1st incomplete Gamma function''}),$$

$$\Gamma(\alpha, x) \equiv \int_x^\infty e^{-t} t^{\alpha-1}\, dt \quad (\text{``2nd incomplete Gamma function''}).$$

$\gamma(\alpha, x)$ and $\Gamma(\alpha, x)$ have the simple series representations

$$\gamma(\alpha, x) = \sum \frac{(-1)^n x^{n+\alpha}}{n!\,(n+\alpha)},$$

$$\Gamma(\alpha, x) = \Gamma(\alpha) - \gamma(\alpha, x). \tag{B.6}$$

(Note further that $\Gamma(n) = (n-1)!$ and $\Gamma(1/2) = \sqrt{\pi}$). The integrals (B.3,4) give the complex Laplace transformation of critical power law decays ($\nu \neq 1$) as well as exponential decays ($\nu = 1$) of the order parameter. A common mean field profile function is $\tanh(x)$ which has a complex Laplace transform of the form

$$\int_0^\infty e^{-\mu x} \tanh(x)\, dx = \beta(\mu/2) - 1/\mu \quad (\text{Re}\{\mu\} > 0). \tag{B.7}$$

$$\beta(\alpha) = \int_0^\infty \frac{t^{\alpha-1}}{1+t}\, dt \quad (\text{Re}\{\alpha\} > 0) \tag{B.8}$$

is the Euler Beta function which can be written as

$$\beta(\alpha) = \sum_{k=0}^\infty (-1)^k / (\alpha + k).$$

The hypergeometric function F and the degenerate hypergeometric function ϕ are defined as

$$F(\alpha; \beta; \gamma; z) = \frac{1}{B(\beta, \gamma - \beta)} \int_0^1 t^{\beta-1} (1-t)^{\gamma-\beta-1} (1-tz)^{-\alpha}\, dt$$

$$(\text{Re}\{\gamma\} > \text{Re}\{\beta\} > 0) \tag{B.9}$$

$$\phi(\alpha; \gamma; z) = \frac{2^{1-\gamma} e^{z/2}}{B(\alpha, \gamma - \alpha)} \int_{-1}^1 (1-t)^{\gamma-\alpha-1} (1+t)^{\alpha-1} \exp\left(\frac{zt}{2}\right) dt$$

$$(\text{Re}\{\gamma\} > \text{Re}\{\alpha\} > 0) \tag{B.10}$$

where $B(x, y)$ is the Euler Beta function of the first kind

$$B(x, y) = \int_0^1 t^{x-1} (1 - t)^{y-1} dt = \frac{\Gamma(x)\,\Gamma(y)}{\Gamma(x + y)}$$

$$(\text{Re}\{x\},\, \text{Re}\{y\} > 0) \quad .$$

$$\text{(B.11)}$$

With the note that $F(-n; \beta; \beta; -z) = (1 + z)^n$ and $\phi(\alpha; \alpha; -z) = e^{-z}$ for arbitrary α, β one immediately follows that $F(1; 1; 1; -z)$ and $\phi(1; 1; -z)$ are a pure Lorentzian and a Gaussian, respectivlely.

Closing Remarks

The *grazing angle x-ray scattering* technique has reached nowadays a very high level of perfection with numerous successful applications in surface science. In two international congresses, one held in Marseille, *"International conference on Surface and Thin Film Studies using Glancing-Angle X-Ray and Neutron Scattering"*, Marseille, France, May 31 – June 2, 1989 [Bienfait and Gay, 1989], the other in Bad Honnef, *"IInd International Conference on Surface X-Ray and Neutron Scattering"*, Bad Honnef, Germany, June 25–28, 1991 [Zabel and Robinson, 1992], the whole body of world-wide activities using this technique became apparent. Yet there still remain many aspects of grazing angle x-ray scattering which have not been explored sufficiently, as time-resolved surface studies which exploit the pulse structure of the synchrotron x-ray beam, as the use of anomalous scattering in order to obtain surface-related partial correlation functions or as grazing angle resonant magnetic x-ray scattering which may allow x-rays to assess interfacial magnetic moments. An experimental challenge which has to be attacked in the near future is the reliable absolute calibration of evanescent x-ray scattering. Only when this can be done satisfactorily, surface-related short range order in binary alloys can be measured in a more quantitative way.

Grazing angle neutron scattering has been discussed in some detail in the workshop on *"Methods of Analysis and Interpretation of Neutron Reflectivity Data"* held in Argonne, Illinois on 23–25 August 1990 [Felcher and Russell, 1991], the next decade will reveal the actual potential of this technique. The moderate surface sensitivity and the comparatively low neutron flux provided by the neutron facilities certainly impose serious experimental limitations. Unfortunately, the further development of this technique (as of other neutron-based research) is jeopardized by the world-wide "neutron draught": Many research reactors have now come into age and have to get upgraded or even be replaced, however, the public acceptance for new nuclear reactors in general is rather low nowadays. On the other hand, from the scientific point of view, neutron spallation sources could be absolutely perfect alternatives to reactors, in addition they would have a quite welcome psychological effect: The neutron scattering community in Europe would finally get a new technology which would entail novel neutron instrumentations. By this one could also overcome to some extent the growing jealousy of the "neutron people" towards the "x-ray people" because of the powerful synchrotron radiation facilities which have been made available to them.

The discussion of the experimental evidence for critical phenomena at surfaces and interfaces may have given the reader the impression, as if the essential features of surface-induced critical effects have been understood nowadays, however, the boot is on the other leg: The surface of a critical system gives rise to a large zoo of new critical exponents of which only one has strictly been confirmed through various experiments in different systems: β_1 which turns out to be universally $\simeq 0.8$ (as tested at the surface of ferromagnets, binary alloys and binary liquids). The two surface-related exponents γ_{11} and $\eta_\parallel$ have been measured so far only in one system and thus there is no experimental evidence for their universality yet. A further critical experimental test of the predictions of the renormalization group theory is the measurement of a critical order parameter profile close to the surface, a very important experiment which has not yet been attempted by the experimentalists. Quite a special case which I want to dwell on are surface critical phenomena related with bulk magnetism in transition metals: On the theoretical side there is an "atomic point of view" which favours quite generally an enhancement of surface magnetic moments, and band theories of magnetism which find quite different answers for different systems. The scarce experimental evidence is very often in some contradiction to the various theories. I want to illustrate this situation with two examples: From a density functional theory one predicts an enhancement of the magnetic moments at the Ni(100) surface from the bulk value $0.56\,\mu_B$ to the surface value $0.68\,\mu_B$ [Wimmer et al., 1984]. This large enhancement of the surface magnetic moments by a factor of 20 % would strongly speak in favour of an extraordinary critical behaviour to occur at this surface, however, SPLEED measurements disclose the ordinary critical exponent β_1 [Alvarado et al., 1982]. The situation is just reversed at the V(100) surface: Here one apparently finds by experiment that V(100) surfaces exhibit ferromagnetic moments in the presence of a paramagnetic bulk, the associated surface Curie temperature being around 540 K [Rau et al., 1986]. These finding are in contradiction to band theories which explicitly don't predict such a surface ferromagnetic phase [Fu et al., 1985]. Independent experimental evidence, as by neutron reflectivity measurements or surface sensitive neutron scattering, would be extremely helpful to shed some new light onto these rather puzzling surface phenomena.

In the course of the investigations of the surface of binary alloys, as Fe_3Al and Cu_3Au, it turned out that the surface-related thermodynamic phenomena are rather complex and by no means fully understood: Even though the general behaviour of the local order parameter near the Cu_3Au (100), Cu_3Au (111) and the Fe_3Al (110) surfaces speaks in favour of a surface-induced disordering which occurs at the bulk transition temperature, there is experimental evidence for an additional order parameter component which originates most likely from segregation effects and survives above the bulk phase transition. The proper understanding of these phenomena requires a systematic study of the composition dependence of this surface-related long-range order.

Finally, todays understanding of surface roughening and surface melting and in particular their interplay is still rather incomplete, however, there is no doubt that grazing-angle x-ray and (may be also) neutron scattering will play a dominant role in providing an unambiguous picture of the microscopic mechanisms leading to these surface disordering phenomena.

References

Abeles, F. [1950]: Ann. Phys. (Paris) **5**, 596
Adair, R.K. [1950]: Rev. Mod. Phys. **22**, 249
Afanase'v, A.M., M.K. Melkonyan [1983]: Acta Cryst. A**39**, 207
Alexander, S. [1987]: in *Transport and Relaxation in Random Materials*, ed. by J. Klafter, R. Rubin, M.F. Schlesinger (World–Scientific, Singapore)
Allen, J.W., S.A. Rice [1977]: J. Chem. Phys. **67**, 5105
Allen, S.M., J.W. Cahn [1976]: Acta Met. **24**, 425
Als–Nielsen, J., O.W. Dietrich [1967]: Phys. Rev. **153**, 706
Als–Nielsen, J., F. Christensen, P.S. Pershan [1982]: Phys. Rev. Lett. **48**, 1107
Als–Nielsen, J. [1986]: Physica A**140**, 376
Al Usta, K., H. Dosch, J. Peisl [1990]: Z. Phys. B**79**, 409
Al Usta, K., H. Dosch, A. Lied, J. Peisl [1991]: Physica B**173**, 65
Al Usta, K., H. Dosch [1991]: private communication
Al Usta, K., H. Dosch, A. Lied, J. Peisl [1992]: in Springer Proceedings in Physics Vol. 61 *"Surface X-Ray and Neutron Scattering"*, ed. by H. Zabel and I.K. Robinson (Springer, Heidelberg)
Alvarado, S.F., M. Campagna, H. Hopster [1982]: Phys. Rev. Lett. **48**, 51
Alvarado, S.F., M. Campagna, A. Fattah, W. Uelhoff [1987]: Z. Phys. B**66**, 103
Andrews, S.R., R.A. Cowley [1985]: J. Phys. C**18**, 6427
Ankner, J.F., H. Zabel, D.A. Neumann, C.F.Majkrzak [1989]: Phys. Rev. B**40**, 792
Ankner, J.F., C.F. Majkrzak, D.A. Neumann, A. Matheny, C.P. Flynn [1991]: Physica B**173**, 89
Ashcroft, N.W., N.D. Mermin [1976]: *Solid State Physics* (Holt–Saunders Int. Editions, New York)

Bacon, G.E. [1975]: *Neutron Diffraction* 3rd ed. (Clarendon Press, Oxford)
Balibar, S., B. Castaign [1985]: Surf. Sci. Rep. **5**, 87
Bardhan, P., J.B. Cohen [1976]: Acta Cryst. A**32**, 597
Batterman, B.W., H. Cole [1964]: Rev. Mod. Phys. **36**, 681
Becker, R.S., J.A. Golovchenko, J.R. Patel [1983]:Phys. Rev. Lett. **50**, 153
Beckmann, P., A. Spizzichino [1963]: *The Scattering of Electromagnetic Waves From Rough Surfaces* (Pergamon Press, New York)
Bernhard, N., E. Burkel, G. Gompper, H. Metzger, J. Peisl, H. Wagner, G. Wallner [1987]: Z. Phys. B**69**, 303
Bethe, H.A., P. Morrison [1956]: *Elementary Nuclear Theory* (J. Wiley, 2nd ed., New York)
Bienenstock, A., J. Lewis [1967]: Phys. Rev. **160**, 393
Bienfait, M., J.M. Gay (eds.) [1990]: *International conference on Surface and Thin Films Studies using Glancing-Angle X-ray and Neutron Scattering*, Journal de Physique Colloque C7
Bilderback, D. [1986]: Nucl. Inst. Met. A**246**, 434
Bilgram, J.H. [1987]: Phys. Rep. **153**, 1
Binder, K., P.C. Hohenberg [1972]: Phys. Rev. B**6**, 3461
Binder, K., P.C. Hohenberg [1974]: Phys. Rev. B**9**, 2194
Binder, K., D. Stauffer [1974]: Phys. Rev. Lett. **33**, 1006
Binder, K. [1983]: in *Phase Transitions and Critical Phenomena*, Vol. 8, ed. by C. Domb, J.L. Lebowitz (Academic Press, New York)
Binder, K., W. Kinzel, W. Selke [1983]: J. Magn. Mat. **31**, 1445

Binder, K., D.P. Landau [1984]: Phys. Rev. Lett. **52**, 318

Binder, K., D.W. Heerman [1985]: in *Scaling Phenomena in Disordered Systems*, ed. by R. Paym, A. Skieltorp (Plenum Press, New York)

Binder, K., D.P. Landau, D.M. Kroll [1986]: Phys. Rev. Lett. **56**, 2272

Birken, H.G., C. Kunz, R. Wolf [1990]: Phys. Script. **41**, 385

Bloch, F. [1936]: Phys. Rev. **50**, 259

Blume, M. [1985]: J. Appl. Phys. **57**, 3615

Bohr, J., R. Feidenhans'l, M. Nielsen, M. Toney, R.L. Johnson [1985]: Phys. Rev. Lett. **54**, 1275

Borelius, G. [1958]: in *Solid State Physics* ed. by F. Seitz and D. Turnbull (Academic Press, New York)

Born, M. [1939]: J. Chem. Phys. **7**, 591

Born, M., E. Wolf [1980]: *Principles of Optics* (Pergamon Press, Oxford)

Borovik–Romanov, A.S., N.M. Kreines, M.A. Talalaev [1971]: JETP Lett. **13**, 54

Braslau, A., M. Deutsch, P.S. Pershan, A.H. Weiss, J. Als–Nielsen, J. Bohr [1985]: Phys. Rev. Lett. **54**, 114

Braslau, A., P.S. Pershan, G. Swislow, B.M. Ocko, J. Als–Nielsen [1987]: Phys. Rev. **A38**, 2457

Bray, A.J., M.A. Moore [1977]: J. Phys. **A10**, 1927; Phys. Rev. Lett. **38**, 1046

Brennan, S., P. Eisenberger [1984]: Nucl. Inst. Meth. **A222**, 164

Broughton, J.Q., G.H. Gilmer [1983]: Act. met. **31**, 845

Brunel, M. [1986]: Acta Cryst. **A42**, 304

Brunel, M., F. de Bergevin [1986]: Acta Cryst. **A42**, 299

Buck, T., G.H. Wheatley, L. Marchut [1983]: Phys. Rev. Lett. **51**, 43

Buffat, P., J.P.Borel [1976]: Phys. Rev. **A13**, 2287

Butler, B.D., J.B. Cohen [1989]: J. Appl. Phys. **65**, 2214

Celotta, R.J., D.T. Pierce, G.C. Wang, S.D. Bader, G.P. Felcher [1979]: Phys. Rev. Lett. **43**, 728

Chen, H., J.B. Cohen, R. Ghosh [1977]: J. Phys. Chem. Sol. **38**, 855

Chernov, A.A., L.V. Mikheev [1988]: Phys. Rev. Lett. **60**, 2488

Chipman, D.R. [1956]: Phys. Rev. **27**, 739

Chui, S.T., J.D. Weeks [1976]: Phys. Rev. **B14**, 4978

Clapp, P.C., S.C. Moss [1968]: Phys. Rev. **171**, 754

Compton, A.H. [1922]: Phys. Rev. **20**, 60

Compton, A.H., K. Allison [1935]: *X-Rays in Theory and Experiment* (D.v.Nostrand, Inc., New York)

Coombes, C.J. [1972]: J. Phys. **F2**, 441

Cotterill, R.M.J., E.J. Jensen, W.D. Kristensen [1974]: in *Anharmonic Lattices, Structural Transitions and Melting*, ed. by T. Riste (Noordhoff Int. Pub., Leiden)

Cowan, P.L. [1985]: Phys. Rev. **B32**, 5437

Cowley, J.M. [1950]: Phys. Rev. **77**, 669; J. Appl. Phys. **21**, 24

Croce, P. [1979]: J. Opt. (Paris) **10**, 141

Dauth, B., W. Duerr, S.F. Alvarado [1987]: Surf. Sci. **189/190**, 729

De Bergevin, F., M. Brunel [1981]: Acta Cryst. **A37**, 314

De Gennes, P. [1985]: Rev. Mod. Phys. **57**, 827

Denier van der Gon, A.W., R.J. Smith, J.M. Gay, D.J. O'Connor, J.F. van der Veen [1990]: Surf. Sci. **227**, 143

Dewames, R.E., T. Wolfram [1969]: Phys. Rev. Lett. **20**, 137

Diehl, H.W. [1983]: in *Phase Transitions and Critical Phenomena*, Vol. **10**, ed. by C. Domb, J.L. Lebowitz (Academic Press, London)

Diehl, H.W., S. Dietrich [1981]: Z. Phys. **B42**, 65; Z. Phys. **B43**, 315

Diehl, H.W., A. Nuesser [1986]: Phys. Rev. Lett. **56**, 2834

Diehl, H.W., A. Nuesser [1990]: Z. Phys. **B79**, 69 and 79

Dietrich, S., H. Wagner [1983]: Phys. Rev. Lett. **51**, 1469

Dietrich, S., H. Wagner [1984]: Z. Phys. **B56**, 207

Dietrich, S., H. Wagner [1985]: Z. Phys. B59, 35
Dietrich, S. [1988]: in *Phase Transitions and Critical Phenomena*, Vol.12, ed. by C. Domb, J.L. Lebowitz (Academic Press, New York)
Ditlevsen, P.D., P. Norskøv, J.K. Stoltze [1991]: (preprint)
Dorner, B. [1972]: Acta Cryst. A28, 319
Dorner, B., R. Comes [1977]: in *Dynamics of Solids and Liquids by Neutron Scattering*, ed. by S.W. Lovesey, T. Springer (Springer Berlin)
Dosch, H., B.W. Batterman, D.C. Wack [1986]: Phys. Rev. Lett. 56, 1144
Dosch, H. [1987]: Phys. Rev. B35, 2137
Dosch, H., L. Mailänder, A. Lied, J. Peisl, F. Grey, R.L. Johnson, S. Krummacher [1988]: Phys. Rev. Lett. 60, 2382
Dosch, H., J. Peisl [1989]: J. de Phys. Colloque C7, 257
Dosch, H., L. Mailänder, H. Reichert, J. Peisl, R.L. Johnson [1991a]: Phys. Rev. B43, 13172
Dosch, H., T. Höfer, J. Peisl, R.L. Johnson [1991b]: Europhys. Lett. 15, 527
Dosch, H., W. Beisler, R.L. Johnson [1991c]: (unpublished results)
Dosch, H., K. Tulipan, G.P. Felcher [1991d]: (unpublished results)
Dürr, H., R. Schneider, T. Fauster [1991]: Phys. Rev. B43, 12187
Du Plessis, J. [1990]: *Surface Segregation*, in *Solid State Phenomena*, Vol. 11 (Sci-Tech Pub., Vaduz)
Dutta, P., S.K. Sinha [1981]: Phys. Rev. Lett. 47, 50

Eckstein, H. [1949]: Phys. Rev. 76, 1328
Eckstein, H. [1950]: Phys. Rev. 78, 731
Eisenberger, P., W.C. Marra [1981]: Phys. Rev. Lett. 46, 1081
Elbaum, M., M. Schick [1991]: Phys. Rev. Lett. 66, 1713
Elson, J.M. [1984]: Phys. Rev. B30, 5460
Epperson, J.E., J.E. Spruiell [1969]: J. Phys. Chem. Sol. 30, 1721 and 1733
Erickson, R.A. [1953]: Phys. Rev. 90, 779

Fankuchen, I. [1938]: Phys. Rev. 53, 910
Faraday, M. [1860]: Proc. Roy. Soc. London 10, 440
Farwig, P., H.W. Schürman [1967]: Z. Phys. 204, 489
Fecht, H.J., W.L. Johnson [1988]: *Nature* 334, 50
Feder, R., H. Pleyer [1982]: Surf. Sci. 117, 285
Feenberg, E. [1932]: Phys. Rev. 40, 40
Feidenhans'l, R. [1989]: Surf. Sci. Rep. 10, No.3
Felcher, G.P. [1981]: Phys. Rev. B24, 1595
Felcher, G.P., R.T. Kampwirth, K.E. Gray, R. Felici [1984]: Phys. Rev. Lett. 52, 1539
Felcher, G.P., R.D. Hilleke, R.K. Crawford, J. Haumann, R. Kleb, G. Ostroski [1987]: Rev. Sci. Inst. 58, 609
Felcher, G.P., T.P. Russell (eds.) [1991]: *Methods of Analysis and Interpretation of Neutron Reflectivity Data*, Physica B173
Felici, R., J. Penfold, R.C. Ward, E. Olsi, C. Matacotta [1987]: Nature 329, 523
Fermi, E., W. Zinn [1946]: Phys. Rev. 70, 103
Fermi, E., L. Marshall [1946]: Phys. Rev. 70, 103
Fisher, M.E., R.J. Burford [1967]: Phys. Rev. 156, 583
Fisher, M.E. [1968]: Phys. Rev. 176, 257
Fisher, M.E. [1971]: in *Critical Phenomena*, ed. by M.S. Green (Academic Press, London)
Fisher, M.E., M.N. Barber [1972]: Phys. Rev. Lett. 28, 1516
Foldy, L.L. [1945]: Phys. Rev. 67, 107
Foldy, L.L. [1955]: Phys. Rev. 83, 688
Frenkel, J. [1955]: *Kinetic Theory of Liquids* (Dover, New York)
Frenken, J.W.M, J.F. van der Veen [1985]: Phys. Rev. Lett. 54, 134
Frenken, J.W.M., J.P. Toennies, C. Wöll [1988]: Phys. Rev. Lett. 60, 1727
Freund, A., G. Marot [1991]: ESRF Newslett. 8, 4

Fu, C.L., A.J. Freeman, T. Oguchi [1985]: Phys. Rev. Lett. **54**, 2700
Fuoss, P.H., I.K. Robinson [1984]: Nucl. Inst. Meth. **A222**, 171
Furukawa, H. [1983]: Phys. Lett. **97A**, 346

Gaulin, B.D., E.D. Hallman, E.C. Svensson [1990]: Phys. Rev. Lett. **64**, 289
Gauthier, Y., R.Baudoing [1989]: in *Surface Segregation and Related Phenomena*, ed. by P.A. Dowben and A. Miller (CRC Press)
Gay, J.M., B. Pluis, J.W.M. Frenken, S. Gierlotka, J.F. van der Veen, J.E. MacDonald, A.A. Williams, W. Piggins, J. Als–Nielsen [1989]: J. de Phys. Colloque C7, 289
Gemmaz, M., M. Afyouni, A. Mosser [1990]: Surf. Sci.Lett. **227**, 109
Gerold, V., K. Kern [1987]: Acta Met. **35**, 393
Gibbons, D.F. [1959]: Phys. Rev. **115**, 1194
Golecki, I., J. Jaccard [1978]: J. Phys. C **11**, 4229
Gompper, G. [1984]: Z. Phys. **B56**, 217
Gompper, G. [1986]: Z. Phys. **B62**, 357
Gompper, G., D.M. Kroll [1988]: Phys. Rev. **B38**, 459
Goos, F., H. Haenchen [1947]: Ann. Phys. **6**, 333
Gorodnichev, E.E., S.L. Dudarev, D.B. Rogozkin, M.I. Ryazanov [1988]: JETP Lett. **48**, 147
Gradshteyn, I.S., I.M. Ryzhik [1965]: *Table of Integrals, Series, and Products* (Academic Press, New York)
Grotehans, S., G. Wallner, E. Burkel, H. Metzger, J. Peisl, H. Wagner [1989]: Phys. Rev. **B39**, 8450
Guentert, O.J. [1965]: J. Appl. Phys. **36**, 1361
Guinier, A. [1963]: *X-Ray Diffraction* (W.H. Freeman and Comp., San Francisco)
Gunton, J.D., M. San Migel, P.S. Sahni [1983]: in *Phase Transitions and Critical Phenomena*, ed. by C. Domb, J.L. Lebowitz (Academic Press, London)
Guttman, L. [1956]: in *Solid State Physics*, Vol. 3, ed. by F. Seitz and D. Turnbull (Academic Press, New York)
Guttman, L., H.C. Schnyders, G.J. Arai [1969]: Phys. Rev. Lett. **22**, 517
Guttman, L., H.C. Schnyders [1969]: Phys. Rev. Lett. **22**, 520

Halperin, B.I., D.R. Nelson [1978]: Phys. Rev. Lett. **41**, 121
Halpern, O. [1949]: Phys. Rev. **75**, 343; Phys. Rev. **76**, 1130
Halpern, O. [1952]: Phys. Rev. **88**, 1003
Hannon, J.P., G.T. Trammell, M. Blume, D. Gibbs [1988]: Phys. Rev. Lett. **61**, 1245
Hansen, M., K. Anderko [1958]: *Constitution of Binary Alloys* (McGraw Hill, New York)
Hayter, J.B., R.R. Highfield, B.J. Pulman, R.K. Thomas, A.I. McMullen, J. Penfold [1981]: J. Chem. Soc. Faraday Trans. 1, **77**, 1437
Heavens, O.S. [1955]: *Optical Properties of Thin Solid Films* (Butterworths Scientific Pub., London)
Helbing, W., B. Dünweg, K. Binder, D.P. Landau [1990]: Z. Phys. **B80**, 401
Held, G.A., J.L. Jordan–Sweet, P.M. Horn, A. Mak, R.J. Birgeneau [1987]: Phys. Rev. Lett. **59**, 2075
Heller, P. [1966]: Phys. Rev. **146**, 403
Helmholtz, H.v. [1886]: Crelles J. **100**, 213
Herring, C. [1952]: in *Structure and Properties of Solid Surfaces*, ed. by R. Gomer and C.S. Smith (University of Chicago Press, Chicago)
Hilliard, J.E., J.W. Cahn [1958]: Acta Met. **6**, 772
Hohenberg, P.C., B.I. Halperin [1977]: Rev. Mod. Phys. **49**, 435
Højlund Nielsen, P.E. [1973]: Phys. Lett. **42A**, 468
Hughes, D.J., M.T. Burgy [1951]: Phys. Rev. **81**, 498
Hughes, D.J. [1954]: *Neutron Optics* (Interscience, New York)

Imry, Y., L. Gunther [1971]: Phys. Rev. **B3**, 3939
Isaacs, E.D., D.B. McWhan. C. Peters, G.E. Ice, D.P. Siddons, J.B. Hastings, C. Vettier, O. Vogt [1989]: Phys. Rev. Lett. **62**, 1671
Israelachvili, J.N. [1985]: *Intermolecular and Surface Forces* (Academic Press, San Diego)

James, R.W. [1948]: *The Optical Principle of the Diffraction of X-rays* (Cornell University Press, Ithaca, New York)

Johnson, R.L., J.H. Fock, I.K. Robinson, J. Bohr, R. Feidenhans'l, J. Als–Nielsen, M. Nielsen, M. Toney [1985]: *The Structure of Surfaces*, ed. by M.A. v.Hove, S.Y. Ton (Springer, Berlin)

Jona, F., J.A. Strozier, C. Wong [1972]: Surf. Sci. **30**, 225

Katano, S., M. Iizumi [1988]: in *Dynamics of Ordering Processes in Condensed Matter*,ed. by S. Komura and H. Furukawa (Plenum Press, New York)

Kauzmann, W. [1948]: Chem. Rev. **43**, 219

Keating, D.T., B.E. Warren [1951]: J. Appl. Phys. **22**, 286

Kendall, K. [1990]: J. Phys. D: Appl. Phys. **23**, 1329

Kiessig, H. [1931]: Ann. Phys. **10**, 715

King, D.A. [1979]: in *Chemistry and Physics of Solid Surfaces*, ed. by R. Vanselow (Chem. Rubber, Cleveland)

Kingetsu, T., M. Yamamoto, S. Nenno [1981]: Surf. Sci. **103**, 13

Koch, E.E. (ed.) [1983]: *Handbook on Synchrotron Radiation*, Vol. 1a,b (North–Holland, Amsterdam)

Köster, L. [1965]: Z. Phys. **182**, 328

Köster, L. [1967]: Z. Phys. **198**, 187

Köster, L., A. Steyerl [1977]: Springer Tracts in Modern Physics, Vol. **80**, (Springer, Heidelberg)

Kosterlitz, J.M., D.J. Thouless [1972]: J. Phys. C5, L124

Kramer, I.R., A. Kumar [1969]: Script. Met. **3**, 205

Kroll, D.M., R. Lipowsky [1983a]: Phys. Rev. B28, 5273 and 6435

Kroll, D.M., G. Gompper [1987]: Phys. Rev. B36, 7078

Kroll, D.M., H. Wagner [1990]: Phys. Rev. B42, 6531

Kuentzler, R. [1973]: Phys. Stat. Sol. **58**

Kumikov, V.K., K.B. Khokonov [1983]: J. Appl. Phys. **54**, 1346

Landolt–Börnstein [1971]: Vol. II (Springer, Berlin)

Lawley, A., J.W. Cahn [1961]: J. Phys. Chem. Sol. **30**, 204

Lax, M. [1950]: Phys. Rev. **80**, 299

Lax, M. [1951]: Rev. Mod. Phys. **23**, 287

Lekner, J. [1991]: Physica B173, 99

Leroux, C., A. Loiseau, M.C. Cadeville, F. Ducastelle [1990a]: Europhys. Lett. **12**, 155

Leroux, C., A. Loiseau, M.C. Cadeville, D. Broddin, G. van Tandeloo [1990b]: J. Phys.: Condens. Matt. **2**, 3479

Levine, J.R., J.B. Cohen, Y.W. Chung, P. Georgopoulos [1989]: J. Appl. Cryst. **22**, 528

Liang, K.S., E.B. Sirota, K.L. D'Amico, G.J. Hughes, S.K. Sinha [1987]: Phys. Rev. Lett. **59**, 2447

Liang, K.S. [1991]: private communication

Lied, A., H. Dosch, L. Mailänder [1989]: (unpublished results)

Lindemann, F.A. [1910]: Z. Phys. **11**, 609

Lipowsky, R [1982]: Phys. Rev. Lett. **49**, 1575

Lipowsky, R., W. Speth [1983]: Phys. Rev. B28, 3983

Lipowsky, R., D.M. Kroll, R.K.P. Zia [1983]: Phys. Rev. B27, 4499

Lipowsky, R., G. Gompper [1984]: Phys. Rev. B29, 5213

Lipowsky, R. [1985]: J. Phys. A18, L585

Lipowsky, R. [1987]: Ferroelectrics **73**, 69

Lipowsky, R., U. Breuer, K.C. Prince, H.P. Bonzel [1989]: Phys. Rev. Lett. **62**, 913

Lloyd, P., M.V. Berry [1967]: Proc. Phys. Soc. London **91**, 678

Löwen, H., T. Beier, H. Wagner [1989]: Europhys. Lett. **9**, 791

Löwen, H., T. Beier [1990]: Phys. Rev. B41, 4435

Löwen, H., R. Lipowsky [1991]: Phys. Rev. B45, 3507

Lovesey, S.W. [1984]: *Theory of Neutron Scattering from Condensed Matter* (Clarendon, Oxford)

Lovesey, S.W. [1987]: J. Phys. C20, 5625

Magerl, A. [1990]: Nucl. Inst. Meth. A290, 414

Maier–Leibnitz, H. [1962]: Z. angew. Phys. **14**, 738

Mailänder, L., H. Dosch, J. Peisl, R.L. Johnson [1990a]: Phys. Rev. Lett. **64**, 2527

Mailänder, L., H. Dosch, J. Peisl, R.L. Johnson [1990b]: *2nd European Conference on Progress in X-Ray Synchrotron Radi+ation Research*, Vol. **25**, ed. by A. Balerna, E. Bernieri, S. Mobilio (Bologna)

Mailänder, L., H. Dosch [1991]: private communication

Mailänder, L., H. Dosch, J. Peisl, R.L. Johnson [1991]: MRS Symp. Proc. **208**, 87

Mandelbrot, B.B. [1982]: *The Fractal Geometry of Nature* (Freeman, New York)

Mansour, A., R.O. Hilleke, G.P. Felcher, R.B. Laibowitz, P. Chaudhari, S.S.P. Parkin [1989]: Physica **B156, 157**, 867

Marra, W.C., P. Eisenberger, A.Y. Cho [1979]: J. Appl. Phys. **50**, 6927

Maradudin, A.A., P.A. Flinn [1963]: Phys. Rev. **129**, 2529

Maradudin, A.A., D.L. Mills [1975]: Phys. Rev. **B11**, 1392

Marshall, W., S.W. Lovesey [1971]: *Theory of Thermal Neutron Scattering* (Clarendon Press, Oxford)

Matsushita, T., U. Kaminaga [1980]: J. Appl. Cryst. **13**, 465 and 472

Mazur, P., D.L. Mills [1982]: Phys. Rev. **B26**, 5175

McDavid, J.M., S.C. Fain Jr. [1975]: Surf. Sci. **52**, 161

McRae, E.G., R.A. Malic [1984]: Surf. Sci. **148**, 551

McRae, E.G., T. Buck [1990]: Surf. Sci. **227**, 67

McReynolds, A.W. [1951a]: Phys. Rev. **83**, 172

McReynolds, A.W. [1951b]: Phys. Rev. **84**, 969

Messiah, A. [1970]: *Quantum Mechanics* (North Holland Pub. Comp., Amsterdam)

Miedema, A.R. [1978]: Z. Metallk. **69**, 287

Mills, D.L. [1975]: Surf. Sci. **48**, 59

Mochrie, S.G.J. [1987]: Phys. Rev. Lett. **59**, 304

Moran–Lopez, J.L., F. Meija–Lira, K.H. Bennemann [1985]: Phys. Rev. Lett. **54**, 1936

Mosher, D., S. Stephanakis [1976]: Appl. Phys. Lett. **29**, 105

Moss, S.C. [1964]: J. Appl. Phys. **35**, 3547

Moss, S.C. [1966]: in *Local Atomic Arrangements Studied by X-Ray Diffraction*, ed. by J.B. Cohen and J.E. Hillard (Gordon and Breach Sci. Pub., New York)

Moss, S.C., P.C. Clapp [1968]: Phys. Rev. **171**, 764

Musket, R.G., W. McLean, C.A. Colmenares, D.M. Makowiecki, W.J. Siekhaus [1982]: Appl. Surf. Sci. **10**, 143

Nagler, S.E., R.F. Shannon, C.R. Harkless, M.A. Singh, R.M. Nicklow [1988]: Phys. Rev. Lett. **61**, 1859

Nagy, E., I. Nagy [1962]: J. Phys. Chem. Sol. **23**, 1605

Nakanishi, H., M.E. Fisher [1983]: J. Chem. Phys. **78**, 3279

Nakanishi, S., K. Kawamoto, N. Fukuoka, K. Umezawa [1992]: Surf. Sci. **261**, 342

Nevot, L., P. Croce [1980]: Rev. Phys. Appl. **15**, 761

Newton, R.G. [1966]: *Scattering Theory of Waves and Particles* (McGraw Hill, New York)

Nishihara, S., Y. Noda, Y. Yamada [1982]: Sol. Stat. Comm. **44**, 1487

Nix, F.C., D. Macnair [1941]: Phys. Rev. **60**, 320

Noda, Y., S. Nishihara, Y. Yamada [1984]: J. Phys. Soc. Jap. **53**, 4241

Ocko, B., A. Braslau, P.S. Pershan, J. Als–Nielsen, M. Deutsch [1987]: Phys. Rev. Lett. **57**, 94

Ohno, K., Y. Okabe, A. Morita [1984]: Prog. Theoret. Phys. **71**, 714

Oshima, K.I., J. Harada, S.C. Moss [1986]: J. Appl. Cryst. **19**, 276

Palmberg, P.W., R.E. Dewames, L.A. Vredevoe [1968]: Phys. Rev. Lett. **21**, 682

Parratt, L.G. [1954]: Phys. Rev. **95**, 359

Penfold, J., R.C. Ward, W.G. Williams [1987]: J. Phys. **E20**, 1411

Penfold, J., R.K. Thomas [1990]: Condens. Mat. **2**, 1369

Pershan, P.S., J. Als–Nielsen [1984]: Phys. Rev. Lett. **52**, 759

Pescia, D., R.F. Willis, J.A.C. Bland [1987]: Surf. Sci. **189**, 724

Picht, J. [1929]: Ann Phys. **5**, 433

Platzman, P.M., N. Tzoar [1970]: Phys. Rev. **B2**, 3556

Pluis, B., T.N. Taylor, D. Frenkel, J.F. van der Veen [1989]: Phys. Rev. **B40**, 1353

Prince, K.C., U. Breuer, H.P. Bonzel [1988]: Phys. Rev. Lett. **60**, 1146

Rabkin, E.I., V.N. Semenov, L.S. Shvindlerman, B.B. Straumal [1991]: Act. met. mat. **39**, 627

Rau, C., C. Liu, A. Schmalzbauer, G. Xing [1986]: Phys. Rev. Lett. **57**, 2311

Reeve, J.S., A.J. Guttman [1980]: Phys. Rev. Lett. **45**, 1581

Reichert, H., H. Dosch, J. Peisl [1992]: Z. Phys. B (to be published)

Reiter, G., S.C. Moss [1986]: Phys. Rev. **B33**, 7209

Ricolleau, C., A. Loiseau, F. Ducastelle [1990]: *3ieme Col. d'expression francaise sur les transitions des phases* (Djerba, 19 − 24 mars 1990)

Rieutord, F. [1990]: Acta Cryst. **A46**, 526

Rottman, C., M. Wortis [1984]: Phys. Rev. **B29**, 328

Roberto, J.B., B.W. Batterman, D.T. Keating [1974]: Phys. Rev. **B9**, 2590

Robinson, I.K. [1983]: Phys. Rev. Lett. **50**, 1145

Robinson, I.K. [1986]: Phys. Rev. **B33**, 3830

Robinson, I.K., E.H. Conrad, D.S. Reed [1990]: J. Phys. France **51**, 103

Robinson, I.K., E. Vlieg, H. Hornis, E.H. Conrad [1991]: Phys. Rev. Lett. **67**, 1890

Russell, T.P., A. Karim, A. Mansour, G.P. Felcher [1989]: Macromol. **21**, 564

Rustichelli, F. [1975]: Phil. Mag. **31**, 1

Rys, F.S. [1986]: Phys. Rev. Lett. **56**, 624

Sakurai, J.J. [1985]: *Modern Quantum Mechanics* (The Benjamin/Cummings Pub. Comp., Menlo Park, CA)

Sanchez, J.M., J.L. Moran–Lopez [1985]: Phys. Rev. **B32**, 3534; Surf. Sci. **157**, L297

Sasaki, Y., K. Hirokawa [1990]: Appl. Phys. **A50**, 397

Saurenbach, F., U. Walz, L. Hinchley, P. Grünberg, W. Zinn [1988]: J. Appl. Phys. **63**, 3473

Schiff, L.I. [1968]: *Quantum Mechanics* (McGraw–Hill, New York)

Schneider, R., L. Belkoura, J. Schelten, D. Woermann, B. Chu [1980]: Phys. Rev. **B22**, 5507

Schulhof, M.P., P. Heller, R. Nathans, A. Linz [1970]: Phys. Rev. Lett. **24**, 1184

Schulhof, M.P., R. Nathans, P. Heller, A. Linz [1971]: Phys. Rev. **B4**, 2254

Schweika, W., K. Binder, D.P. Landau [1990]: Phys. Rev. Lett. **65**, 3321

Schwinger, J. [1937]: Phys. Rev. **51**, 544

Schwinger, J. [1948]: Phys. Rev. **73**, 407

Scott, J.M. [1933]: *The Land that God Gave Cain* (Chatto & Windus, London)

Sears, V.F. [1982]: Phys. Rep. **82**, 1

Sears, V.F. [1986]: Phys. Rep. **141**, 281

Sears, V.F. [1989]: *Neutron Optics* (Oxford University Press, New York)

Shaler, A.J. [1953]: in *Structure and Properties of Solid Surfaces*, ed. by R. Gomer and C.S. Smith (University of Chicago Press, Chicago)

Shapira, Y., C.C. Becerra [1976]: Phys. Lett. **54A**, 483

Shinjo, T. [1991]: Surf. Sci. Rep. **12**, 50

Sigl, L., W. Fenzl [1986]: Phys. Rev. Lett. **57**, 2191

Sinha, S.K., E.B. Sirota, S. Garoff, H.B. Stanley [1988]:Phys. Rev. **B38**, 2297

Smoluchowski, R. [1944]: Phys. Rev. **60**, 661

Snell, A.H. [1973]: in *A Random Walk in Science* (ed. by E. Mendoza, The Institute of Physics, London)

Sommerfeld, A. [1972]: *Partial Differential Equations in Physics* (Academic Press, New York)

Sparks, C.J., B.S. Borie, J.B. Hastings [1980]: Nucl. Inst. Meth. **172**, 237

Spencer, J., H. Winick [1980]: *"Wiggler Systems as Sources of Electromagnetic Radiation"* in *Synchrotron Radiation Research*, ed. by H. Winick and S. Doniach (Plenum Press, New York)

Squires, G.L. [1978]: *Introduction to the Theory of Thermal Neutron Scattering* (Cambridge University Press, Cambridge)

Stamm, M., G. Reiter, S. Hüttenbach [1989]: Physica B156 & 157, 564
Stern, E.A., Z. Kalman, A. Levis, K. Liebermann [1988]: Appl. Opt. 27, 5135
Steyerl, A. [1972]: Z. Phys. 254, 169
Stock, K.D. [1980]: Surf. Sci. 91, 655
Stoltze, P., J.K. Norskøv, U. Landmann [1988]: Phys. Rev. Lett. 61, 440
Stoltze, P., J.K. Norskøv, U. Landmann [1989]: Surf. Sci. 220, L693
Stoner, E.C. [1929]: Phil. Mag. 8, 250
Stroud, D., N.W. Ashcroft [1972]: Phys. Rev. B5, 371
Sundaram, V.S., B. Farrell, R.S. Alben, W.D. Robertson [1973]: Phys. Rev. Lett. 31, 1136
Sundaram, V.S., R.S. Alben, W.D. Robertson [1974]: Surf. Sci. 46, 653

Teaney, D.T. [1965]: Phys. Rev. Lett. 14, 898
Thiel, D.J., E.A. Stern, D.H. Bilderback, A. Lewis [1989]: Physica B158, 314
Trayanov, A., E. Tosatti [1988]: Phys. Rev. B38, 6961
Trayanov, A., A.C. Levi, E. Tosatti [1989]: Europhys. Lett. 8, 657
Trümper, J. [1990]: Phys. Bl. 46, 137
Trümper, J. [1991]: Phys. Bl. 47, 29

Uelhoff, W. [1987]: Festkörperprobleme 27, 241

van Beijeren, H. [1977]: Phys. Rev. Lett. 38, 993
van der Veen, J.F., B. Pluis, A.W. Denier van der Gon [1988]: in Chemistry and Physics of Solid Surfaces VII, Springer Series Surf. Sci. 10, ed. by R. Vanselow and R.F. Howe (Springer, Heidelberg)
Varshni, Y.P. [1970]: Phys. Rev. B2, 3952
Vaughan, D. (ed.) [1986]: X-Ray Data Booklet (Lawrence Berkeley Lab., Berkeley)
Vetterling, W.T., R.V. Pound [1976]: J. Opt. Soc. Am. 66, 1048
Vineyard, G.H. [1982]: Phys. Rev. B26, 4146
Vidal, B., P. Vincent [1984]: Appl. Opt. 23, 1794
Villain, J., D.R. Grempel, J. Lapujoulade [1985]: J. Phys. F15, 809
Vlieg, E., A.v. Ent, A.P. Jongh, H. Neerings, J.F. van der Veen [1987]: Nucl. Inst. Meth. A262, 522
Voges, D., E. Taglauer, H. Dosch, J. Peisl [1992]: Surf. Sci. (accept. for publ.)
von Blanckenhagen, P., W. Schommers, V. Voegele [1987]: J. Vac. Sci. Tech. A5, 649
Voronel, A., S. Rabinovich, A. Kisliuk, V. Steinberg, T. Sverbilova [1988]: Phys. Rev. Lett. 60, 2402

Wagner, H. [1984]: (unpublished results)
Wagner, H. [1985]: in Applications of Field Theory to Statistical Mechanics, ed. by L. Garrido, Lecture Notes in Physics, Vol. 216 (Springer, Heidelberg)
Warlimont, H. (ed.) [1974]: Order-Disorder Transformations in Alloys (Springer, Heidelberg)
Warner, M., J.E. Gubernatis [1985]: Phys. Rev. B32, 6347
Warren, B.E., D.R. Chipman [1949]: Phys. Rev. 75, 1629
Warren, B.E., J.S. Clarke [1965]: J. Appl. Phys. 36, 324
Warren, B.E. [1969]: X-Ray Diffraction (Addison–Wesley, Reading, MA)
Watanabe, M., T. Hidaka, H. Tanino, K. Hoh, Y. Mitsuhashi [1984]: Appl. Phys. Lett. 45, 725
Weber, W., B. Lengeler [1991]: (preprint)
Weeks, J.D. [1980]: in Ordering in Strongly Fluctuating Condensed Matter Systems, ed. by T. Riste (Plenum Press, New York)
Weil, N.A., S.A. Bortz, R.F. Firestone [1963]: Mat. Sci. Res. 1, 291
Widom, B. [1978]: J. Chem. Phys. 68, 3878
Wimmer, E., A.J. Freeman, H. Krakauer [1984]: Phys. Rev. B30, 3113
Woodruff, D.P. [1973]: The Solid-Liquid Interface (Cambridge Univ. Press, London)
Wright, C.S. [1924]: Miscellaneous Data (Scott Polar Expedition of 1911–1912) comp. by H.G. Lyons (Harrison and Sons, London)

Yoneda, Y. [1963]: Phys. Rev. 131, 2010
Yun, W.B., J.M. Bloch [1990]: J. Appl. Phys. 68, 1421

Zabel, H., I.K. Robinson [1992]: *"Surface X-Ray and Neutron Scattering"*, Springer Proc. in Physics, Vol. 61 (Springer, Heidelberg)

Zachariasen, W.H. [1967]: *Theory of X-Ray Diffraction in Crystals* (Dover Pub. Inc., New York)

Zeilinger, A., T.J. Beatty [1983]: Phys. Rev. B27, 7239

Zeppenfeld, P., K. Kern, R. David, G. Comsa [1989]: Phys. Rev. Lett. 62, 63

Zhu, X.M., R. Feidenhans'l, H. Zabel, J. Als–Nielsen, R. Du, C.P. Flynn, F. Grey [1988]: Phys. Rev. B37, 7157

Zhu, X.M., I.K. Robinson, E. Vlieg, H. Zabel, J.A. Dura, C.P. Flynn [1989]: J. Phys. Colloque C7, 283

Zhu, X.M., H. Zabel, I.K. Robinson, E. Vlieg, J.A. Dura, C.P. Flynn [1990]: Phys. Rev. Lett. 65, 2692

Subject Index

N. G. Chetaev

Theoretical Mechanics

Translated from the Russian by I. Aleksanova

1989. 407 pp. 190 figs. Hardcover
ISBN 3-540-51379-5

This university-level textbook reflects the extensive teaching experience of N. G. Chataev, one of the most influential teachers of theoretical mechanics in the Soviet Union. The mathematically rigorous presentation largely follows the traditional approach, supplemented by material not covered in most other books on the subject. To stimulate active learning numerous carefully selected exercises are provided. Attention is drawn to historical pitfalls and errors that have led to physical misconceptions.

Extensive appendices contain material from additional lectures on optics and mechnics analogies, Poincaré's equation and the special theory of elasticity.

Distribution rights for the socialist countries, India and Iran:
V/O "Mezhdunarodnaya Kniga", Moscow

D. Park, Williams College, Williamstown, MA

Classical Dynamics and Its Quantum Analogues

2nd enl. and updated ed. 1990. IX, 333 pp. 101 figs. Hardcover ISBN 3-540-51398-1

The primary purpose of this textbook is to introduce students to the principles of classical dynamics of particles, rigid bodies, and continuous systems while showing their relevance to subjects of contemporary interest. Two of these subjects are quantum mechanics and general relativity. The book shows in many examples the relations between quantum and classical mechanics and uses classical methods to derive most of the observational tests of general relativity. A third area of current interest is in nonlinear systems, and there are discussions of instability and of the geometrical methods used to study chaotic behaviour. In the belief that it is most important at this stage of a student's education to develop clear conceptual understanding, the mathematics is for the most part kept rather simple and traditional.

This book devotes some space to important transitions in dynamics: the development of analytical methods in the 18th century and the invention of quantum mechanics.

A. Hasegawa, AT & T Bell Laboratories, Murray Hill, NJ

Optical Solitons in Fibers

2nd enl. ed. 1990. XII, 79 pp. 25 figs.
Softcover ISBN 3-540-51747-2

Already after six months high demand made a new edition of this textbook necessary. The most recent developments associated with two topical and very important theoretical and practical subjects are combined: **Solitons** as analytical solutions of nonlinear partial differential equations and as lossless signals in dielectric **fibers.** The practical implications point towards technological advances allowing for an economic and undistorted propagation of signals revolutionizing telecommunications. Starting from an elementary level readily accessible to undergraduates, this pioneer in the field provides a clear and up-to-date exposition of the prominent aspects of the theoretical background and most recent experimental results in this new and rapidly evolving branch of science. This well-written book makes not just easy reading for the researcher but also for the interested physicist, mathematician, and engineer. It is well suited for undergraduate or graduate lecture courses.

Springer-Verlag
Berlin
Heidelberg
New York
London
Paris
Tokyo
Hong Kong
Barcelona
Budapest

A. G. Sitenko, Academy of the Ukrainian SSR

Scattering Theory

1991. XI, 294 pp. 32 figs. (Springer Series in Nuclear and Particle Physics)
Hardcover ISBN 3-540-51953-X

This book is an introduction to nonrelativistic scattering theory. The presentation is mathematically rigorous, but is accessible to upper level undergraduates in physics. The relationship between the scattering matrix and physical observables, i. e. transition probabilities, is discussed in detail. Among the emphasized topics are the stationary formulation of the scattering problem, the inverse scattering problem, dispersion relations, three-particle bound states and their scattering, collisions of particles with spin and polarization phenomena. The analytical properties of the scattering matrix are discussed. Problems round off this volume.

B. N. Zakhariev, Moscow; **A. A. Suzko,** Minsk, USSR

Direct and Inverse Problems

Potentials in Quantum Scattering

1990. XIII, 223 pp. 42 figs.
Softcover ISBN 3-540-52484-3

This textbook can almost be viewed as a "how-to" manual for solving quantum inverse problems, that is, for deriving the potential from spectra or scattering data and also, as somewhat of a quantum "picture book" which should enhance the reader's quantum intuition. The formal exposition of inverse methods is paralleled by a discussion of the direct problem. Differential and finite-difference equations are presented side by side. The common features and (dis)advantages of a variety of solution methods are analyzed. To foster a better understanding, the physical meaning of the mathematical quantities are discussed explicitly. Wave confinement in continuum bound states, resonance and collective tunneling, energy shifts and the spectral and phase equivalence of various interactions are some of the physical problems covered.

K. L. G. Heyde, University of Gent, Belgium

The Nuclear Shell Model

1990. XII, 376 pp. 171 figs. (Springer Series in Nuclear and Particle Physics) Hardcover
ISBN 3-540-51581-X

This book evolved from a course in theoretical nuclear physics taught over seven years at the University of Gent and is thus well suited to and tested for lecture courses. The nuclear shell model is introduced from basic techniques such as angular momentum and tensor algebra. The material is developed from the beginning up to the present state-of-the-art calculations using self-consistent residual interactions. Problem sets and simple computer codes are included to facilitate a better acquaintance with the subject.
The appendices constitute an integral part of the text going into depth on a number of technical derivations to provide the reader with a detailed background facilitating active research. The book introduces the subject to advanced undergraduate and to graduate students providing them with knowledge and techniques for own research in this field. It is a highly useful prerequisite for lecturers teaching modern nuclear physics.

Springer-Verlag
Berlin
Heidelberg
New York
London
Paris
Tokyo
Hong Kong
Barcelona
Budapest

Springer Tracts in Modern Physics

---

* denotes a volume which contains a Classified Index starting from Volume 36